GTO
RED BOOK

Pontiac GTO 1964–1974

Peter C. Sessler

*Special thanks to Bart R. Orlans, Gene Marmos,
Jim Mattison, the GTO Association of America and Pontiac
Historical Society*

First published in 1992 by Motorbooks International Publishers & Wholesalers, PO Box 2, 729 Prospect Avenue, Osceola, WI 54020 USA

© Peter C. Sessler

All rights reserved. With the exception of quoting brief passages for the purposes of review no part of this publication may be reproduced without prior written permission from the Publisher

Motorbooks International is a certified trademark, registered with the United States Patent Office

The information in this book is true and complete to the best of our knowledge. All recommendations are made without any guarantee on the part of the author or Publisher, who also disclaim any liability incurred in connection with the use of this data or specific details

We recognize that some words, model names and designations, for example, mentioned herein are the property of the trademark holder. We use them for identification purposes only. This is not an official publication

Motorbooks International books are also available at discounts in bulk quantity for industrial or sales-promotional use. For details write to Special Sales Manager at the Publisher's address

Library of Congress Cataloging-in-Publication Data Available
ISBN 0-87938-611-8

On the front cover: The gold 1967 Pontiac GTO owned by Duffy Cummings of Amarillo, Texas. *Jerry Heasley*

All photos by the author except as indicated

Printed and bound in the United States of America

Contents

	Introduction	5
Chapter 1	**1964 GTO**	7
Chapter 2	**1965 GTO**	13
Chapter 3	**1966 GTO**	20
Chapter 4	**1967 GTO**	27
Chapter 5	**1968 GTO**	35
Chapter 6	**1969 GTO**	43
Chapter 7	**1970 GTO**	52
Chapter 8	**1971 GTO**	61
Chapter 9	**1972 GTO**	67
Chapter 10	**1973 GTO**	73
Chapter 11	**1974 GTO**	79
	Appendix	85

Introduction

This book is designed to help the Pontiac enthusiast determine the authenticity and originality of any GTO built between 1964 and 1974. Each chapter covers a model year. Included are production figures; serial number decoding information; engine block codes; carburetor, distributor and cylinder head numbers; option order codes and retail prices; color and trim codes; and selected facts.

If one car epitomizes the sixties muscle car phenomenon, that car is the Pontiac GTO. Like the Mustang, which was also introduced in 1964, the GTO happened to be the right car at the right time. The GTO concept—install a large passenger car engine in an intermediate body platform—was fairly simple, but its success was due to good timing: the public was ready for a smaller performance car. Within a year, emulators abounded, with other General Motors divisions offering their own versions of the GTO, but the original outsold them all.

For the enthusiast, the most important number in any GTO is its vehicle identification number (VIN). In 1964, it consisted of an eight-digit number that broke down to number of cylinders, model series, model year, assembly plant code and consecutive sequence number. From 1965 to 1971, it consisted of a thirteen-digit number that broke down to model number, model year, assembly plant and consecutive sequence number. In 1972, Pontiac (as well as the rest of GM) adopted a new VIN numbering system. The new system still used thirteen digits, but most important, it included a code identifying the engine the car was equipped with. Until 1971, you could not tell which engine came in a particular car.

The car's VIN was stamped on a metal plate and attached to the driver's-side door pillar until 1967. From 1968 on, it was stamped on a plate and attached to the left side of the dash, making it visible through the windshield.

Pontiac also stamped the last eight digits of the car's VIN on a pad on the right-hand front of the engine block, thereby "matching" the engine to the car. Unfortunately, it is easy to restamp a cylinder block with the VIN, thereby increasing a car's value. (See the Appendix for further information designed to help you determine authenticity.)

Even if all the numbers match on a particular car you are looking at, especially on one built before 1972, it would be to your advantage if the car is documented. It would be all the better if the previous owner can provide you with the original invoice or window sticker, any service records, a Protect-O-Plate or the car's broadcast sheet. The broadcast sheet, usually located under the seats or carpets or behind the dash, is a factory computer printout showing what options the GTO was built with.

The colors and interior trim listed in each chapter are correct as far as they go. However, Pontiac did build cars in colors and trim combinations not listed. As with all the information listed here, be open to the possibility that exceptions can and do occur. This means that you'll have to work harder to determine authenticity.

Although every effort has been made to make sure that the information contained in this book is correct, I cannot assume any responsibility for any loss arising from the use of this book. However, I would like to hear from any enthusiast with any corrections or interesting additions. Please write to me care of Motorbooks International.

Chapter 1

1964 GTO

Production

By Body Style
2 dr coupe	7,384
2 dr hardtop	18,422
2 dr convertible	6,644
Total	32,450

By Engine
389 ci 4 bbl V-8	24,205
389 ci 3 × 2 bbl V-8	8,245
Total	32,450

Serial Numbers

Description
824P100001
8—Number of cylinders
2—Model series (2–LeMans)
4—Last digit of model year (4–1964)
P—Assembly plant (P–Pontiac, M–Kansas City, F–Freemont, B–Baltimore)
100001—Consecutive sequence number

Location
On plate attached to left front door hinge post.

Engine Block Codes
78X, 78XX—389 ci 325 hp 3 speed manual
78W, 78XW—389 ci 325 hp 4 speed wide-ratio manual
78X9, 78W9—389 ci 325 hp 4 speed wide-ratio manual, 3.90:1 gears
79J—389 ci 325 hp 2 speed automatic
76X, 76XX—389 ci 348 hp three speed manual
76W, 76XW—389 ci 348 hp 4 speed wide-ratio manual

Engine Block Codes
76X9, 76W9—389 ci 348 hp 4 speed wide-ratio manual, 3.90:1 gears
77J—389 ci 348 hp 2 speed automatic

Carburetors
389 ci 325 hp manual—3647S
389 ci 325 hp automatic—3649S
389 ci 348 hp manual—7024178 front, 7024175 center, 7024179 rear
389 ci 348 hp automatic—7024178 front, 7024173 center, 7024179 rear

Distributors
Conventional ignition—1111054
Transistor ignition—1111057

Cylinder Head Casting Number
389 ci, all—9770716

Option Order Codes and Retail Prices

2227 LeMans sports coupe	$2,491.00
2237 LeMans hardtop coupe	2,556.00
2367 LeMans convertible	2,796.00
389 ci 3 × 2 bbl engine	115.78
4 speed manual transmission	188.30
2 speed automatic transmission	199.06
382 GTO option	295.90
392 Push-button radio w/manual antenna	
Coupe	62.41
Convertible	88.77
393 Push-button radio w/electric antenna	
Coupe	92.16
Convertible	118.52
394 AM/FM radio w/manual antenna	150.54
395 AM/FM radio w/electric antenna	180.39
401 Sepera-Phonic rear speaker	14.15
402 Dual-speed electric wipers	4.84
404 Underhood lamp	3.55
414 Seatbelt deletion	(11.00)
421 Dual-speed windshield washer & wipers	17.27
424 Instrument panel pad	16.14

431 HD air cleaner	4.84
441 Visor vanity mirror	1.45
442 Nonglare inside mirror	4.25
444 Remote control LH outside mirror	11.78
451 Remote control deck lid	10.76
452 Tachometer	53.80
454 Tilt steering wheel	43.04
462 Deluxe wheel discs	17.22
471 Back-up lamps	12.64
474 Verbra-Phonic rear speaker	53.80
481 Luggage lamp	3.55
491 Courtesy lamp	4.30
501 Power steering	96.84
502 Wonder Touch power brakes	42.50
504 Rally gauge cluster & tachometer	86.08
512 Door edge guards	4.84
524 Custom Sports steering wheel	39.27
531 Soft Ray glass, all windows	31.20
541 Rear window defogger	21.52
551 Power windows	106.25
564 Power seat, LH bucket	71.02
572 Spare wheel tire cover	2.58
581 Tri-Comfort AC	345.60
582 HD 61 amp battery	3.55
584 Heater deletion	(73.00)
602 Outside rearview mirror	4.25
604 Electric clock	19.37
612 Handling Kit	16.14
614 Positive crankcase ventilation	5.38
622 Superlift rear shocks	40.35
624 Custom retractable front seatbelts	7.53
631 1 pair front floor mats	6.24
632 1 pair rear floor mats	5.81
671 Transistor ignition	75.27
692 Metallic brake linings	36.86
701 Safe-T-Track differential	37.66
Solid paint, special colors	40.19
Two-tone paint	31.74
Two-tone paint, special colors	65.93

Exterior Colors
Starlight Black   A

Exterior Colors
Cameo Ivory	C
Silvermist Grey	D
Yorktown Blue	F
Skyline Blue	H
Pinehurst Green	J
Marimba Red	L
Sunfire Red	N
Aquamarine	P
Gulfstream Aqua	Q
Alamo Beige	R
Saddle Bronze	S
Singapore Gold	T
Grenadier Red	V
Nocturne Blue	W

Interior Colors
Black	214
Blue	215
Saddle	216
Aqua	217
Red	218
Parchment	219

Convertible Top Colors
Ivory	1
Black	2
Blue	4

Facts

The GTO option was available on three body styles on the LeMans series: sports coupe, hardtop coupe and convertible. They were all two-door models.

Exterior identification was through the use of GTO lettering on the blacked-out left front part of the grille, on the right rear deck lid and on the rear quarter panels. A GTO 6.5 liter crest was mounted on each front fender, behind the front wheelwell. The hood came with two simulated hood scoops.

In the interior, a GTO emblem was located above the glovebox. Other interior features standard on the GTO were bucket seats, an engine-turned appliqué on the instrument panel and a Hurst shifter on manual-transmission-equipped cars. Some early cars got Hurst shifters without the familiar Hurst inscription on the shift lever. All interiors except Parchment ones came with matching door panels and trim; with Parchment interiors, the seats, door panels, headliner, sun visors and windlace were Parchment and the carpets, console base, rear package shelf, door kick panels and dash pad were black.

The four-pod instrument panel contained the following gauges and indicator lamps: oil pressure indicator lamp and battery, far left; 120 mph speedometer and odometer, left center; fuel gauge and temperature indicator lamp, right center; and a blank pod, far right. If the optional 7000 rpm tachometer or the Rally clock was ordered, it would be located in the far right pod. Also optional was a brake warning lamp, located at the bottom of the speedometer. An optional manifold vacuum gauge was available, as well; it was mounted on the console.

Two steering wheels were available. The standard GTO wheel was the LeMans Deluxe wheel, which matched the interior in color, except with Parchment interiors, which got a black wheel. The optional Custom Sports steering wheel came with four brushed metal spokes and an imitation wood rim.

All GTO radios could be had with either the standard manual antenna, mounted on the front cowl, or the optional power antenna, located on the rear quarter panel.

The standard GTO engine was the 389 ci V-8 rated at 325 hp with a 10.75:1 compression ratio. It came with a Carter four-barrel carburetor. Optional was the 389 ci 348 hp V-8 using the Rochester Tri-Power triple two-barrel carburetors. The end carburetors were vacuum controlled. All 389 ci V-8s came with chrome valve covers and a chrome oil filler cap.

The standard transmission was a three-speed manual with a Hurst floor shifter. Optional were two four-speeds: the M20 wide-ratio box and the M21 close ratio. The two-speed Powerglide automatic was optional on both engines.

All GTOs came with a dual-exhaust system. Optional with this were chrome exhaust splitters located behind the rear wheelwells. Early splitters were shorter than later production splitters.

Manual steering with a 24:1 ratio and five turns lock to lock and manual four-wheel drum brakes were standard equipment. Power assist was optional. Drum brakes with metallic shoes were optional. The suspension came with firmer shocks and springs and a 0.938 in. front anti-sway bar.

Standard wheels measured 14 × 6 in. with 7.50 × 14 Red Line tires. Whitewalls were a no-cost option. Standard with the wheel was a hubcap with a ten-slot Deluxe wheel cover; an eight-slot Custom wheel cover and a wire wheel cover were optional.

1964 Pontiac Le Mans GTO.

Chapter 2

1965 GTO

Production

By Body Style
2 dr coupe	8,319
2 dr hardtop	55,722
2 dr convertible	11,311
Total	75,352

By Engine
389 ci 4 bbl V-8	54,805
389 ci 3 × 2 bbl V-8	20,547
Total	75,352

By Transmission
Manual	56,378
Automatic	18,974
Total	75,352

Serial Numbers

Description
237275P100001
23727—Model number (23727–2 dr coupe, 23737–2 dr hardtop, 23767–2 dr convertible)
5—Last digit of model year (5–1965)
P—Assembly plant (A–Atlanta, P–Pontiac, B–Baltimore, K–Kansas City, R–Arlington, Z–Freemont)
100001—Consecutive sequence number

Location
On plate attached to left front door hinge post.

Engine Block Codes
WT—389 ci 4 bbl 335 hp manual
YS—389 ci 4 bbl 335 hp automatic
WS—389 ci 3 × 2 bbl 360 hp manual
YR—389 ci 3 × 2 bbl 360 hp automatic

Carburetors
389 ci 335 hp manual—3895S
389 ci 335 hp automatic—3896S
389 ci 360 hp manual—7024178 front, 7025175 center, 7024179 rear
389 ci 360 hp automatic—7024178 front, 7025177 center, 7024179 rear

Distributors
389 ci 335 hp—1111054
389 ci 335 hp, transistor ignition—1111080
389 ci 360 hp, transistor ignition—1111047

Cylinder Head Casting Number
389 ci—77

Option Order Codes and Retail Prices
23727 2 dr LeMans coupe	$2,787.00
23737 2 dr LeMans hardtop	2,852.00
23767 2 dr LeMans convertible	3,093.00
389 ci 3 × 2 bbl engine	115.78
2 speed automatic transmission	199.06
342 Black Cordova top	75.32
346 Fawn Cordova top	75.32
382 GTO option	295.90
392 AM push-button radio w/manual antenna	62.41
393 AM push-button radio w/electric antenna	92.16
394 AM/FM push-button radio w/manual antenna	136.65
395 AM/FM push-button radio w/electric antenna	166.40
398 AM manual control radio w/manual antenna	53.80
399 AM manual control radio w/electric antenna	83.55
401 Sepera-Phonic rear speaker	14.15
404 Underhood lamp	3.55
414 Seatbelt deletion	(11.00)
421 Dual-speed windshield washer & wipers	17.27

424 Instrument panel pad	16.14
432 HD radiator wo/582	15.06
441 Visor vanity mirror	1.45
442 Nonglare inside mirror	4.25
444 Remote control LH outside mirror	11.78
451 Remote control deck lid	10.76
454 Tilt steering wheel	43.04
462 Deluxe wheel disc	17.22
471 Back-up lamps	
W/manual transmission	12.91
W/automatic transmission	10.76
474 Verbra-Phonic rear speaker	53.80
481 Luggage lamps	3.55
482 Glovebox lamps	2.85
492 Ashtray & lighter lamps	3.12
494 Parking brake lamp	4.95
501 Power steering	96.84
502 Power brakes	42.50
504 Rally gauge cluster & tachometer	86.08
512 Door edge guards	4.84
524 Custom Sports steering wheel	39.27
531 Soft Ray glass, all windows	31.20
532 Soft Ray glass, windshield only	19.91
541 Rear window defogger	21.52
551 Power windows	102.22
564 Power seat, LH bucket only	71.02
582 Tri-Comfort AC	345.60
584 Heater deletion	(73.00)
601 Console	48.15
602 Outside rearview mirror	4.25
604 Electric clock	19.37
612 20:1 quick-ratio steering	10.76
614 Positive crankcase ventilation	5.38
621 Ride & Handling Package	16.14
622 Superlift rear shocks	40.35
634 Safeguard speedometer & fuel warning lamp	16.14
641 Rally wheels	52.72
662 Transistorized voltage regulator	10.76
671 Transistor ignition	
W/582	75.27
Wo/582	64.51

Option Order Codes and Retail Prices

701 Safe-T-Track differential	37.66
Two-tone paint, standard colors	31.74
Two-tone paint, special colors	71.93
Solid paint, special color	40.19

Exterior Colors

Starlight Black	A
Blue Charcoal	B
Cameo Ivory	C
Fontaine Blue	D
Nightwatch Blue	E
Palmetto Green	H
Reef Turquoise	K
Teal Turquoise	L
Burgundy	N
Irish Mist	P
Montero Red	R
Capri Gold	T
Mission Beige	V
Bluemist Slate	W
Mayfair Maize	Y

Interior Colors

Black	214-30
Blue	217-33
Turquoise	214-36
Gold	215-34
Red	216-35
Parchment and Black	218-3E

Convertible Top Colors

White	1
Black	2
Blue	4
Turquoise	5
Beige	6

Vinyl Top Colors

Black	2
Beige	6

Facts

The 1965 LeMans was restyled and so too was the GTO, which was still an option package on the two-door coupe, hardtop and convertible. The front end of the car had a totally new look with vertical headlights and a split, recessed grille. The hood came with a single, centrally located nonfunctional hood scoop. At the rear, the most noticeable change was the wraparound taillights. GTO emblems were again located on the left front grille, on the rear quarter panels and on the deck lid. The location of the 6.5 liter crest remained unchanged: behind the front wheelwells. An additional side embellishment was a thin pinstripe running the length of the car.

Complementing the new wraparound taillights was a new rear panel, which ran the width of the rear. The name Pontiac was cast in the panel, which had six horizontal ribs (the regular Tempest panel had ten ribs). The six ribs on the GTO matched the ribs on the taillight lenses.

The trunk was painted spatter blue with green and gray flecks, whereas the 1964 trunks were painted body color.

The interior was revised only slightly. Instead of the engine-turned dash appliqué, a wood veneer was used. This could be either vinyl or real wood. Although the dash pad was similar to the 1964 version, the gray on it was different. Other changes included the addition of a vinyl-trimmed assist bar located over the glovebox door and the use of a different GTO emblem located to the left of the assist bar.

The four-pod instrument cluster remained unchanged. However, a new Rally gauge cluster with different instruments was an option. In the far left pod, the fuel gauge was located on the top, with the battery indicator lamp on the bottom. In the left center pod was the 120 mph speedometer and the odometer, with a parking brake indicator light beneath the odometer optional. An unfurled checkered flag was located just above the center of the speedometer. In the right center pod was an 8000 rpm tachometer. Two types were made: the early cars came with a green band that swept from 0 to 8000 rpm, and later cars had the green turn to red from 5200 to 8000 rpm. The far right pod contained a 0°F to 230°F water temperature gauge on the top and a 0-30-60 lb oil pressure gauge. The Rally clock was optional with the standard instrument cluster. The heater control panel was redesigned, with a movable vent located above it.

The standard steering wheel, again the LeMans Deluxe wheel, came with a round horn ring. The optional Custom Sports steering wheel was a deep, three-spoke design.

An additional radio was made available: a mono AM/FM, which could be had with either the standard manual antenna or the optional electric antenna.

A console was optional with either manual or automatic transmissions. It was the same unit used in 1964, but the areas between the chrome ribs were painted gloss-black. The seat upholstery was changed to a diagonal rib pattern. A small GTO 6.5 liter crest was mounted on the front top part of the door panels.

Mechanically, the standard 389 ci single four-barrel V-8 was rated at 335 hp and the optional Tri-Power 389 ci V-8 put out 360 hp. The increase in power was due to different camshafts and improved intake manifolds. The linkage on the Tri-Power engines remained vacuum controlled on automatic-equipped GTOs, whereas GTOs with a manual transmission used a mechanical, progressive linkage. A late 1965 release was the dealer- or owner-installed Air Scoop Package, which made the standard hood scoop functional. Tri-Power GTOs with manual transmission came with larger, 2.2 in. exhaust tailpipes, versus the 2 in. pipes used for all other GTOs. Exhaust resonators were built into the tailpipes, with exhaust splitters optional.

The compression ratio, at 10.75:1, was unchanged.

GTOs destined for California came with a positive crankcase ventilation (PCV) emission system. Four-barrel cars came with a black snorkel-type air cleaner with a chrome lid, whereas the forty-nine-state cars got an open pancake-type air cleaner. All engines came with chrome valve covers and a chrome oil filler cap.

Transmission availability was unchanged from that in 1964. From March 1965 on, however, the three-speed was replaced by a Ford three-speed gearbox.

The only noteworthy suspension change was the use of cross-member reinforcement brackets on GTOs equipped with a manual transmission. This was instituted in June of 1965.

In addition to the optional ten-rib Deluxe wheel covers, the Custom wheel covers (which came with a simulated knock-off spinner) and the wire wheel covers (which got cooling slots for 1965), new optional Rally wheels were available. These were stamped steel wheels with five cooling slots painted in silver, and they came with a trim ring and a center cap. All wheels measured 14 × 6 in., but tires were increased to 7.75 × 14 Red Lines, with whitewalls a no-cost option.

1965 Pontiac GTO convertible.

Chapter 3

1966 GTO

Production

By Body Style
2 dr coupe	10,363
2 dr hardtop	73,785
2 dr convertible	12,798
Total	96,946

By Engine
389 ci 4 bbl	77,901
389 ci 3 × 2 bbl	19,045
Total	96,946

By Transmission
Manual	61,279
Automatic	35,667
Total	96,946

Serial Numbers

Description
242076P100001
24207—Model number (24207–2 dr coupe, 24217–2 dr hardtop, 24267–convertible)
6—Last digit of model year (6–1966)
P—Assembly plant (P–Pontiac, B–Baltimore, G–Framingham, K–Kansas City, Z–Freemont)
100001—Consecutive sequence number

Location
On plate attached to left front door hinge post.

Engine Block Codes
WT—389 ci V-8 4 bbl 335 hp manual
WW—389 ci V-8 4 bbl 335 hp manual w/AIR
YS—389 ci V-8 4 bbl 335 hp automatic
XE—389 ci V-8 4 bbl 335 hp automatic w/AIR
WS—389 ci V-8 3 × 2 bbl 360 hp manual
WV—389 ci V-8 3 × 2 bbl 360 hp manual w/AIR

XS—389 ci V-8 3 × 2 bbl 360 hp manual Ram Air
YR—389 ci V-8 3 × 2 bbl 360 hp automatic

Carburetors
389 ci 335 hp manual—4033S
389 ci 335 hp manual w/AIR—4041S
389 ci 335 hp automatic—4034S
389 ci 335 hp automatic w/AIR—4030S
389 ci 360 hp manual Ram Air—7025178 front, 7026075 center, 7025179 rear
389 ci 360 hp manual w/AIR—7025178 front, 7036175 center, 7025179 rear
389 ci 360 hp automatic—7024178 front, 7026074 center, 7024179 rear

Distributors
389 ci 335 hp, conventional ignition—1111078
389 ci 360 hp, conventional ignition—1111054
389 ci 335 or 360 hp, transistor ignition—1111047
389 ci 335 or 360 hp, manual w/AIR—1111103

Cylinder Head Casting Number
389 ci, all—093

Option Order Codes and Retail Prices

2 door coupe	$2,783.00
2 dr hardtop	2,847.00
2 dr convertible	3,082.00
291 White Cordova top	84.26
292 Black Cordova top	84.26
296 Fawn Cordova top	84.26
342 AM push-button radio w/manual antenna	61.09
343 AM push-button radio w/electric antenna	90.21
344 AM/FM push-button radio w/manual antenna	133.76
345 AM/FM push-button radio w/electric antenna	162.88
348 AM manual control radio w/manual antenna	52.66

Option Order Codes and Retail Prices

349 AM manual control radio w/electric antenna	81.78
351 Sepera-Phonic rear speaker	15.80
352 Verbra-Phonic rear speaker	52.66
372 Spare wheel tire cover	2.53
374 Rear window defogger	21.06
382 Door edge guards	4.74
391 Visor vanity mirror	1.68
392 Nonglare inside mirror	4.16
394 Remote control LH outside mirror	7.37
401 Luggage lamp	3.48
402 Glovebox lamp	2.79
404 Roof, rail & reading lamps	13.59
411 Instrument panel courtesy lamp	4.21
412 Ashtray & lighter lamps	3.05
414 Parking brake lamp	4.85
421 Underhood lamp	3.48
422 Remote control deck lid	12.64
431 Custom front & rear seatbelts	10.53
438 1 pair front seat shoulder harnesses	26.33
441 Safeguard speedometer	15.80
444 Electric clock	18.96
452 Wire wheel discs	69.58
454 Rally wheels	56.80
458 Custom wheel discs	36.33
461 Deluxe wheel discs	16.85
471 Custom Sports steering wheel	38.44
472 Console	47.13
482 Tailpipe extensions	30.23
501 Power steering	94.79
502 Power brakes	41.60
504 Tilt steering wheel	42.13
511 20:1 quick-ratio steering	10.53
514 HD 7 blade fan & clutch	3.05
521 Traffic hazard warning flasher switch	11.59
522 Red fender liners	26.33
524 Walnut 4 speed gearshift knob	3.69
531 Soft Ray glass, all windows	30.23
532 Soft Ray glass, windshield only	19.49
551 Power windows	100.05
564 Power seat, LH bucket only	69.51

571 RH & LH headrests	52.66
574 RH reclining seats & RH & LH headrests	84.26
582 Custom AC	343.20
584 Heater deletion	(71.76)
614 Positive crankcase ventilation	5.27
621 Ride & Handling Package	3.74
628 Soft Ride springs & shocks	NC
631 1 pair front floor mats	6.11
632 1 pair rear floor mats	5.69
634 Superlift rear shocks	39.50
651 HD aluminum front brake drums	49.08
662 Transistorized voltage regulator wo/582	10.53
671 Transistor ignition	
W/582	73.67
Wo/582	63.14
681 HD radiator wo/582	14.74
692 HD 55 amp alternator	15.80
731 Safe-T-Track differential	36.86
782 Automatic transmission	194.84
784 4 speed manual transmission	184.31
802 Trophy 389 3 × 2 bbl engine	113.33
Two-tone paint, standard colors	31.07
Two-tone paint, special colors	114.27
Solid paint, special color	83.20

Exterior Colors

Starlight Black	A
Blue Charcoal	B
Cameo Ivory	C
Fontaine Blue	D
Nightwatch Blue	E
Palmetto Green	H
Reef Turquoise	K
Marina Turquoise	L
Burgundy	N
Barrier Blue	P
Montero Red	R
Martinique Bronze	T
Mission Beige	V
Platinum	W
Candlelite Cream	Y

Interior Colors
Blue	219
Turquoise	220
Bronze	221
Red	222
Black	223
Parchment	224

Convertible Top Colors
White	1
Black	2
Blue	4
Turquoise	5
Beige	6

Vinyl Top Colors
Ivory	1
Black	2
Beige	6

Facts

The A-body platform was totally redesigned for 1966. Still using a 115 in. wheelbase, the GTO was longer and heavier, though it resembled the 1965 model from the front. It became a separate model line, which still included the three body styles: hardtop, coupe and convertible.

The split grille used vertical headlights, as before, but the grille openings tapered at the top and bottom as the grille approached the center dividing panel. The turn signal lamps were relocated next to the headlights and were attached to the plastic grille.

Plastic was used for the first time on the grilles. The hood, with its simulated scoop, was identical with the 1965's, except for the Pontiac crest located on the front center. From the side, the 1966 had a more distinct "Coke bottle" look. All 1966s used a thin side pinstripe, rocker panel moldings that continued along the bottom of the rear fenders and wheelwell moldings. GTO emblems were located in the same places as before: on the grille, on the rear deck and on the rear fenders. The GTO 6.5 liter crest was located on the front fenders behind the wheelwells.

The trunk was painted with blue and green spatter paint and came with a green hound's-tooth-pattern floor mat.

Taillight treatment, too, was redesigned. The single taillight lamp unit was covered by a housing that had three horizontal louvers.

Although the dash panel was redesigned, its look was evolutionary, as it utilized the familiar four-pod theme. Instrument and gauge selection was the same as in 1965. Optional on the standard panel were the Rally clock and the Safeguard speedometer. The Safeguard speedometer, which included a low-fuel warning lamp, allowed the driver to use a pointer to select a speed that when exceeded would sound a warning buzzer.

The vent found above the heater and air conditioner controls on the 1965 GTOs was included on the 1966s only if air conditioning was ordered. A vinyl-trimmed assist bar was located above the glovebox door, with a GTO nameplate located to its left. The instrument panel was covered with wood veneer.

The standard steering wheel was a two-spoke Deluxe unit, and the optional Custom Sports steering wheel was unchanged from 1965.

Although bucket seats were standard equipment, an optional Strato Bench front seat was available.

The standard three-speed manual transmission came with a column-mounted shifter. Hurst shifters were standard equipment on the heavy-duty three-speed and four-speed manual boxes. The close-ratio four-speed was available only with 3.9:1 and 4.33:1 rear axle ratios.

Engine availability was unchanged, with a 389 ci 335 hp V-8 standard and the 360 hp Tri-Power optional. The Tri-Power came with a larger center carburetor. During the model year, another engine option was released. This was known as the XS engine option, after its engine block code. It included the Air Scoop Package and a revised camshaft, and it was the first Ram Air engine. The Air Scoop Package was also revised to accommodate the larger center carburetor. The compression ratio was 10.75:1 on all engines.

A set of 9.5 in. drum brakes was still standard fare on the GTO, with metallic linings optional. An aluminum front drum option was part of the heavy-duty Ride and Handling Package. Exhaust splitters were not used; instead, chrome exhaust extensions exiting beneath the rear bumper were optional. Exhaust pipes and tailpipes measured 2 in. with all engines.

The standard hubcap was a carry-over, as were the Deluxe wheel covers, but the Pontiac Motor Division lettering was deleted. The wire wheel cover was similar to the one used in 1965, but it incorporated a two-spoke spinner. New for 1966 was the Custom wheel cover, which had twelve chrome and black spokes. The handsome Rally wheel was unchanged except that the center hubcap was painted flat-black around the lug nuts. All wheels measured 14×6 in., and 7.75×14 Red Line tires were once again standard, with whitewalls a no-cost option.

This was the first year that the red fender liners were used. These were plastic shells molded to fit both front and rear wheelhouses.

1966 Pontiac GTO convertible.

Chapter 4

1967 GTO

Production

By Body Style
2 dr coupe	7,029
2 dr hardtop	65,176
2 dr convertible	9,517
Total	81,722

By Engine
400 ci 255 hp	2,967
400 ci 335 hp	64,177
400 ci 360 hp HO	13,827
400 ci 360 hp Ram Air	751
Total	81,722

By Transmission
Manual	39,128
Automatic	42,594
Total	81,722

Serial Numbers

Description
242077P100001
24207—Model number (24207–2 dr coupe, 24217 –2 dr hardtop, 24267–2 dr convertible)
7—Last digit of model year (7–1967)
P—Assembly plant (P–Pontiac, B–Baltimore, G–Framingham, K–Kansas City, Z–Freemont)
100001—Consecutive sequence number

Location
On plate attached to left front door hinge post.

Engine Block Codes
XM—400 ci V-8 2 bbl 255 hp automatic
XL—400 ci V-8 2 bbl 255 hp automatic w/AIR
WT—400 ci V-8 4 bbl 335 hp manual

Engine Block Codes
WW—400 ci V-8 4 bbl 335 hp manual w/AIR
YS—400 ci V-8 4 bbl 335 hp automatic
XE—400 ci V-8 4 bbl 335 hp automatic w/AIR
WS—400 ci V-8 4 bbl 360 hp manual
WV—400 ci V-8 4 bbl 360 hp manual w/AIR
YZ—400 ci V-8 4 bbl 360 hp automatic
YI—400 ci V-8 4 bbl 360 hp automatic w/AIR
XS—400 ci V-8 4 bbl 360 hp manual Ram Air
YR—400 ci V-8 4 bbl 360 hp manual Ram Air w/AIR
XP—400 ci V-8 4 bbl 360 hp automatic Ram Air
YQ—400 ci V-8 4 bbl 360 hp automatic Ram Air w/AIR

Carburetors
400 ci 255 hp—7027060
400 ci 255 hp w/AIR—7037092
400 ci 335 hp—7027063
400 ci 335 hp w/AIR—7037263
400 ci 335 hp automatic—7027262
400 ci 335 hp automatic w/AIR—7037263
400 ci 360 hp—7027263 (incl Ram Air)
400 ci 360 hp w/AIR—7037263 (incl Ram Air)
400 ci 360 hp automatic—7027262 (incl Ram Air)
400 ci 360 hp automatic w/AIR—7037262 (incl Ram Air)

Distributors
400 ci 255 hp—1111242
400 ci 335 hp—1111244
400 ci 360 hp—1111254
400 ci, transistor ignition—1111251

Cylinder Head Casting Numbers
400 ci 255 hp—142
400 ci 335 and 360 hp—670; 97 for Ram Air after serial number 646616

Option Order Codes and Retail Prices
2 dr coupe	$2,871.00
2 dr hardtop	2,935.00
2 dr convertible	3,165.00

341 Power antenna	29.12
342 AM push-button radio w/manual antenna	61.09
343 AM push-button radio w/electric antenna	90.58
344 AM/FM push-button radio w/manual antenna	133.76
345 AM/FM push-button radio w/electric antenna	162.88
348 AM manual control radio w/manual antenna	52.66
349 AM manual control radio w/electric antenna	81.78
351 Sepera-Phonic rear speaker	15.80
352 Verbra-Phonic rear speaker	52.66
354 Stereo tape deck	128.00
372 Spare wheel tire cover	2.53
374 Rear window defogger	21.06
382 Door edge guards	4.74
391 Visor vanity mirror	1.68
392 Nonglare inside mirror	4.16
394 Remote control LH outside mirror	7.37
401 Luggage lamp	3.48
402 Glovebox lamp	2.79
404 Roof, rail & reading lamps	3.59
411 Instrument panel courtesy lamp	4.21
412 Ashtray & lighter lamps	3.05
414 Parking brake lamp	4.85
421 Underhood lamp	3.48
422 Remote control deck lid	12.64
431 Front & rear seatbelts	10.51
444 Rally gauge cluster & hood tachometer	84.26
452 Wire wheel discs	69.58
453 Rally II wheels	73.00
454 Rally I wheels	56.86
458 Custom wheel discs	36.33
461 Deluxe wheel discs	16.85
471 Wood-grain steering wheel	30.02
472 Console	68.46
482 Tailpipe extensions	30.23
501 Power steering	94.79
504 Tilt steering wheel	42.13
511 20:1 quick-ratio steering	10.53
514 HD 7 blade fan & clutch	3.05

Option Order Codes and Retail Prices

521 Power front disc brakes	104.79
522 Red fender liners	26.33
524 Walnut 4 speed gearshift knob	3.69
531 Soft Ray glass, all windows	30.23
532 Soft Ray glass, windshield only	19.49
551 Power windows	100.05
564 Power seat, LH bucket only	69.51
568 Bench front seat	NC
571 RH & LH headrests	52.66
578 RH reclining bucket seat	84.26
582 Custom AC	343.20
584 Heater deletion	(71.76)
614 Positive crankcase ventilation	5.27
621 Ride & Handling Package	3.74
628 Soft Ride springs & shocks	NC
631 1 pair front floor mats	6.11
632 1 pair rear floor mats	5.69
634 Superlift rear shocks	39.50
651 HD aluminum front brake drums	49.08
662 Transistorized voltage regulator wo/582	10.53
671 Ignition capacitor discharge	104.26
681 HD radiator wo/582	14.74
692 HD 55 amp alternator	15.80
731 HD rear end differential	63.19
782 Automatic transmission	226.44
784 4 speed manual transmission	184.31
400 ci 255 hp economy engine	NC
400 ci 360 hp engine	76.89
Ram Air Package	263.30
Cordova top	84.26
Two-tone paint, standard colors	31.07
Two-tone paint, special colors	114.27
Solid paint, special color	83.20

Exterior Colors

Starlight Black	A
Cameo Ivory	C
Montreaux Blue	D
Fathom Blue	E
Tyrol Blue	F
Signet Gold	G

Linden Green	H
Gulf Turquoise	K
Mariner Turquoise	L
Plum Mist	M
Burgundy	N
Silverglaze	P
Regimental Red	R
Champagne	S
Montego Cream	T

Interior Colors

	Strato Buckets	Bench
Blue	219	—
Turquoise	220	—
Gold	221	—
Black	223	235
Parchment	224	236
Red	225	—

Convertible Top Colors

Ivory-White	1
Black	2
Blue	4
Turquoise	5
Cream	7

Vinyl Top Colors

Ivory	1
Black	2
Cream	7

Facts

The 1967 was essentially a carry-over in terms of styling, but with differences. The plastic grille insert was replaced by a chrome wire mesh design. The rear taillight panel treatment was new, using four horizontal slots on each side. The side profile was the same as in 1966, but the rocker panel molding was much larger and incorporated the familiar GTO 6.5 liter crest. GTO emblems were located on the grille, on the deck lid and on each rear fender.

In the interior, vinyl was used to simulate wood on the dash. The appearance of the dash panel was the same; however, the 1967 unit was not interchangeable with the 1966 unit. One easy way to differentiate the two is that the 1967 dash panel used a split oval turn signal indicator. The standard instrument package was unchanged, but the optional Rally cluster had some minor revisions. The water temperature gauge went to 250°F, and the oil pressure gauge read to 80 lb. Some early Rally clusters were carry-over 1966 units. Also, a few Rally clusters got the water and oil gauges in the right center pod and a Rally clock in the far right pod. This cluster arrangement was available when the hood-mounted tachometer was ordered, but it could be had by itself as well. The redline band began at 5100 rpm on the hood tachometer, which had white numerals on a steel-blue background.

Radio availability was complemented with a stereo eight-track player. It was mounted under the dash beneath the radio.

Other interior changes included a different seat pattern comprising three sets of three ribs. This was complemented by matching horizontal ribs on the door panels. A notchback bench seat was optional. The steering wheel design was changed as well. The standard wheel was a three-spoker that incorporated the horn buttons on the spokes. The center ornament had the Pontiac crest and was marked Energy Absorbing to let the driver know that the car came with an energy-absorbing steering column. The optional Custom Sports steering wheel, also a three-spoke design, was smaller in diameter than the 1966 unit and had a larger hub.

In addition to the Hurst floor shifters used on manual-transmission-equipped cars, the Hurst Dual Gate shifter (which allowed the driver to shift through the gears) was used on GTOs equipped with the new three-speed Turbo Hydra-matic automatic. The Turbo Hydra-matic replaced the two-speed Powerglide automatic.

Four engines were available. The standard GTO engine displaced 400 ci and was rated at 335 hp. A no-cost option was the 400 ci 255 hp V-8, which came with a two-barrel carburetor and an 8.6:1 compression ratio. Optional were two more 400 ci V-8s, both rated at 360 hp. The 400 High Output (HO) used a single four-barrel Rochester carburetor, an open-element air cleaner, free-flowing exhaust manifolds and a camshaft that had more lift.

The 400 Ram Air came with the HO's open-element air cleaner but also included all the parts necessary to make the stock nonfunctional hood scoop functional. The air pan and open scoop were shipped in the trunk for the dealer to install.

A small number of Tri-Power engines were built and installed on early 1967 GTOs before a corporate ban on multiple carburetion took effect.

Engines destined for California got the Air Injection Reactor (AIR) emission system, which consisted primarily of an air pump and related hardware. These engines came with a black snorkel air cleaner that had a chrome lid.

Dual exhausts were standard on all GTOs. Manual-transmission-equipped cars had 2.25 in. tailpipes, whereas cars equipped with the Turbo Hydra-matic automatic had 2 in. pipes and 22 in. long resonators.

The braking system was upgraded. A federally mandated dual master cylinder was used on all GTOs. The standard drum brakes still measured 9.5 in., but for the first time, power front disc brakes, measuring 11.5 in., were optional.

An additional styled steel wheel became available: the Rally II five-spoke wheel. This was painted gray, with the raised portion on each spoke painted argent. The center cap was red. The Rally II wheel used on cars with the optional disc brakes had a different offset to clear the disc brake caliper. In addition, the Rally I wheel was still available, unchanged from 1966. Three wheel covers were also available: the Deluxe cover with its six cutouts; the Custom cover, also with six cutouts but including a three-prong spinner; and a wire wheel cover. Standard tires were F70×14 Red Lines, with whitewalls a no-cost option.

1967 Pontiac GTO.

Chapter 5

1968 GTO

Production

By Body Style
2 dr hardtop	77,704
2 dr convertible	9,980
Total	87,684

By Engine and Transmission

Hardtop
400 ci 265 hp automatic	2,841
400 ci 350 hp manual	25,371
400 ci 350 hp automatic	39,215
400 ci 360 hp manual	6,197
400 ci 360 hp automatic	3,140
400 ci 360/366 hp manual Ram Air	757
400 ci 360/366 hp automatic Ram Air	183
Total	77,704

Convertible
400 ci 265 hp automatic	432
400 ci 350 hp manual	3,116
400 ci 350 hp automatic	5,091
400 ci 360 hp manual	766
400 ci 360 hp automatic	461
400 ci 360/366 hp manual Ram Air	92
400 ci 360/366 hp automatic Ram Air	22
Total	9,980

Serial Numbers

Description
242378P100001
24237—Model number (24237–2 dr hardtop, 24267–2 dr convertible)
8—Last digit of model year (8–1968)
P—Assembly plant (A–Atlanta, B–Baltimore, G–Framingham, P–Pontiac, R–Arlington,

Z–Freemont)
100001—Consecutive sequence number

Location
On plate attached to driver's side of dash, visible through the windshield.

Engine Block Codes
XM—400 ci V-8 2 bbl 265 hp automatic
WT—400 ci V-8 4 bbl 350 hp manual
YS—400 ci V-8 4 bbl 350 hp automatic
WS—400 ci V-8 4 bbl 360 hp manual HO
YZ—400 ci V-8 4 bbl 360 hp automatic HO
XS—400 ci V-8 4 bbl 360 hp manual Ram Air I
XP—400 ci V-8 4 bbl 360 hp automatic Ram Air I
WY—400 ci V-8 4 bbl 366 hp manual Ram Air II
XW—400 ci V-8 4 bbl 366 hp automatic Ram Air II

Carburetors
400 ci 265 hp—7028060
400 ci 350 hp manual—7028263
400 ci 350 hp automatic—7028268
400 ci 360 hp manual—7028267 (HO)
400 ci 360 hp automatic—7028268 (HO)
400 ci 360 hp manual—7028275 (Ram Air I)
400 ci 360 hp automatic—7028274 (Ram Air I)
400 ci 366 hp manual—7028273 (Ram Air II)
400 ci 366 hp automatic—7028270 (Ram Air II)

Distributors
400 ci 265 hp—1111272
400 ci 4 bbl manual—1111449
400 ci 4 bbl automatic—1111270, 1111300

Cylinder Head Casting Numbers
400 ci 265 hp—14
400 ci 350/360 hp HO—16
400 ci 360 hp Ram Air I—31
400 ci 366 hp Ram Air II—R96A

Option Order Codes and Retail Prices
2 dr hardtop	$3,101.00
2 dr convertible	3,327.00

321 Basic Group	89.53
322 Protection Group	
Hardtop	55.22
Convertible	34.16
331 Mirror Group	
Hardtop	13.69
Convertible	9.48
332 Lamp Group	5.27
346 400 ci 2 bbl engine w/351	NC
347 400 ci 4 bbl Ram Air engine	342.29
348 400 ci 4 bbl HO engine	76.88
351 Turbo Hydra-matic automatic transmission	236.97
354 4 speed manual transmission	184.31
358 Close-ratio 4 speed manual transmission	184.31
361 Safe-T-Track differential	
Exc HD	42.13
HD	63.19
382 AM push-button radio w/manual antenna	61.09
383 AM push-button radio w/electric antenna	90.58
384 AM/FM push-button radio w/manual antenna	113.76
385 AM/FM push-button radio w/electric antenna	163.25
388 AM/FM stereo radio w/manual antenna	239.08
389 AM/FM stereo radio w/electric antenna	268.57
391 Sepera-Phonic rear speaker	15.80
392 Verbra-Phonic rear speaker	52.66
394 Stereo tape player	133.76
402 Spare wheel tire cover	5.27
404 Rear window defogger	21.06
412 Door edge guards	6.24
414 Concealed headlamps	52.66
421 RH visor vanity mirror	2.11
422 RH & LH visor vanity mirrors	4.21
424 Remote control LH outside mirror	9.48
431 Custom front & rear seatbelts	9.48
432 Rear shoulder belts, exc convertible	26.33
434 Hood-mounted tachometer	63.19
441 Cruise control	52.66
444 Rally gauge cluster & tachometer (NA w/442 or 484)	84.26

Option Order Codes and Retail Prices

452 Wire wheel discs	73.72
452 Wire wheel discs	69.58
453 Rally II wheels	84.26
454 Rally I wheels	61.09
458 Custom wheel discs	41.07
461 Deluxe wheel discs	21.06
471 Custom Sports steering wheel	30.54
472 Console	
W/manual transmission	52.66
W/automatic transmission	68.46
482 Tailpipe extensions	21.06
484 Rally gauge cluster & clock	50.55
492 Remote control deck lid	13.69
494 Rally stripes	10.53
501 Power steering	94.79
502 Power brakes	42.13
504 Tilt steering wheel	42.13
514 HD 7 blade fan & clutch	5.27
521 Disc brakes, front only	63.19
524 Gearshift knob, w/floor shift	4.21
531 Soft Ray glass, all windows	34.76
532 Soft Ray glass, windshield only	25.28
551 Power windows	100.05
564 Power seat, LH bucket only	69.51
571 RH & LH head restraints	52.66
578 RH reclining bucket seat w/RH & LH head restraints	84.26
582 Custom AC	360.20
621 Ride & Handling Package	4.21
622 HD springs & shocks	4.21
631 1 pair front floor mats	6.85
632 1 pair rear floor mats	6.32
634 Superlift rear shocks	42.13
651 Cornering lamps	33.70
652 Luggage lamp	3.16
671 Underhood lamp	3.16
672 Ignition switch lamp	2.11
692 HD 55 amp alternator	15.80
702 Space Saver spare tire	15.80
731 Dual-stage air cleaner	9.48
732 Rear compartment throw mat	8.43

754 Front shoulder belts
W/431	26.33
Wo/431	23.17
Cordova (vinyl) top	94.79
Solid paint, special color	83.20
Two-tone paint, standard colors	40.02
Two-tone paint, special colors	123.22

Exterior Colors

Starlight Black	A
Cameo Ivory	C
Alpine Blue	D
Aegena Blue	E
Nordic Blue	F
April Gold	G
Autumn Bronze	I
Meridian Turquoise	K
Aleutian Blue	L
Flambeau Burgundy	N
Springmist Green	P
Verdoro Green	Q
Solar Red	R
Primavira Beige	T
Nightshade Green	V
Mayfair Maize	Y

Interior Colors

	Buckets	**Bench**
Teal	219	—
Turquoise	220	—
Gold	221	—
Black	223	235
Parchment	224	236
Red	225	—

Convertible Top Colors

Ivory-White	1
Black	2
Teal	5
Gold	8

Vinyl Top Colors
Ivory	1
Black	2
Teal	5
Gold	8

Facts

The 1968 GTO was totally new as far as the exterior and interior went. The chassis and drivetrain were basically unchanged. Most noticeable was the Endura front bumper, which was color keyed to the car's paint. Optional was a conventional chrome bumper, the same unit used on the LeMans. Also optional was a hidden headlamp option, which was vacuum actuated. The hood used two simulated scoops, and it was designed to hide the windshield wipers.

A GTO emblem was used on the left front grille opening and on the right rear part of the deck lid. Replacing the emblems on the rear fenders were GTO decals. The GTO 6.5 liter crest was located behind the front wheelwell on the front fenders.

The trunk was painted in dark blue and gray spatter paint and came with a dark gray one-piece mat.

In the interior, bucket seats were standard fare. The instrument panel was redesigned using three pods. On the standard panel, the left pod housed the fuel gauge and warning lights for oil pressure, alternator and water temperature functions. The center pod was for the 120 mph speedometer and odometer, which also included a brake warning lamp. The Safeguard speedometer was optional with the standard panel. The right pod was where the optional Rally clock was located. Two Rally clusters were available, one with an 8000 rpm tachometer and the other with the Rally clock. Both of these were located in the right pod. On either cluster, the left-hand pod contained the fuel gauge, oil pressure gauge, water temperature gauge and generator warning lamp. The bezels on the control knobs used either black (early) or white lettering to describe their functions. Simulated woodgrain appliqué was used on the instrument panel.

A GTO crest was used on both door panels.

Three hood tachometers were used. The first type (installed until mid-December 1967) and the second type (installed afterwards) differed only with respect to the division lines between the rpm numerals. The later tachometer had longer lines. Both indicated a 5200 rpm redline. A third tachometer was used exclusively on GTOs equipped with the Ram Air II engine. It showed a 5500 rpm redline.

Another radio was added to the option list. This was an AM/FM stereo radio. Along with the stereo eight-track player, this radio used a 6×9 in. rear shelf-mounted speaker with the 4×10 in. front unit to create the stereo effect.

The standard steering wheel was the same unit used in 1967, except that the horn bars matched the car's interior color. The optional Custom Sports steering wheel was smaller and shallower.

A Hurst shifter was used on all three- and four-speed manual transmissions. The Hurst Dual Gate shifter was optional with the Turbo Hydra-matic automatic. Otherwise, the automatic's shift lever was located on the steering column.

The standard engine was the 400 ci V-8 rated at 350 hp. The economy 400 ci V-8 was available only with the automatic transmission and was rated at 265 hp. The 400 HO, rated at 360 hp, was the first performance engine option, followed by the 400 Ram Air, also rated at 360 hp. The Ram Air engine was replaced by the Ram Air II engine in March of 1968. The Ram Air II came with forged pistons, a forged steel crankshaft and new cylinder heads that featured round exhaust ports. Not available with air conditioning, it was rated at 366 hp and could be had with either the M21 close-ratio four-speed manual or the three-speed automatic. Only one axle ratio was available, 4.33:1, and the Safe-T-Track limited slip was a mandatory option. Also included was a Ram Air induction system, which was to be installed by the selling dealer. The system was not driver controlled, but rather the driver was expected to replace the open scoops with the standard nonfunctional scoops during bad weather.

The brakes were a carry-over from 1967, as were the front and rear suspensions. On the exhaust system, automatic-equipped cars got a 2 in. tailpipe and manual-equipped cars came with larger 2.25 in. tailpipes. Chrome exhaust extensions were optional.

Rally I and Rally II wheels were optional. These were unchanged with the exception of the Rally II's lug nuts, which had black inserts. Three other wheel covers were available: the Deluxe wheel covers, which were a radial spoke design; the Custom wheel covers, which came with eight round openings; and the wire wheel covers, which were a carry-over from 1967. All wheels measured 14×6 in.

In an industrywide move, the GTO's VIN plate was relocated to the left front part of the dash panel and was visible through the windshield.

1968 Pontiac GTO convertible.

Chapter 6

1969 GTO

Production

By Body Style
2 dr GTO hardtop	58,126
2 dr GTO convertible	7,328
2 dr The Judge hardtop	6,725
2 dr The Judge convertible	108
Total	72,287

By Engine and Transmission

GTO Hardtop & The Judge Hardtop
400 ci 265 hp automatic	1,246
400 ci 350 hp manual	22,032
400 ci 350 hp automatic	32,744
400 ci 366 hp manual Ram Air III	6,143
400 ci 366 hp automatic Ram Air III	1,986
400 ci 370 hp manual Ram Air IV	549
400 ci 370 hp automatic Ram Air IV	151
Total	64,851

GTO Convertible & The Judge Convertible
400 ci 265 hp automatic	215
400 ci 350 hp manual	2,415
400 ci 350 hp automatic	4,385
400 ci 366 hp manual Ram Air III	249
400 ci 366 hp automatic Ram Air III	113
400 ci 370 hp manual Ram Air IV	45
400 ci 370 hp automatic Ram Air IV	14
Total	7,436

The Judge Hardtop
400 ci 366 hp manual Ram Air III	4,894
400 ci 366 hp automatic Ram Air III	1,534
400 ci 370 hp manual Ram Air IV	239
400 ci 370 hp automatic Ram Air IV	58
Total	6,725

The Judge Convertible

400 ci 366 hp manual Ram Air III	74
400 ci 366 hp automatic Ram Air III	29
400 ci 370 hp manual Ram Air IV	5
Total	108

Serial Numbers

Description
242379A100001
24237—Model number (24237–2 dr hardtop, 24267–2 dr convertible)
9—Last digit of model year (9–1969)
A—Assembly plant (A–Atlanta, B–Baltimore, G–Framingham, P–Pontiac, R–Arlington, Z–Freemont)
100001—Consecutive sequence number

Location
On plate attached to driver's side of dash, visible through the windshield.

Engine Block Codes
XX—400 ci V-8 2 bbl 265 hp manual
XM—400 ci V-8 2 bbl 265 hp automatic
WT—400 ci V-8 4 bbl 350 hp manual
YS—400 ci V-8 4 bbl 350 hp automatic
WS—400 ci V-8 4 bbl 366 hp manual Ram Air III
YZ—400 ci V-8 4 bbl 366 hp automatic Ram Air III
WW—400 ci V-8 4 bbl 370 hp manual Ram Air IV
XP—400 ci V-8 4 bbl 370 hp automatic Ram Air IV

Carburetors
400 ci 265 hp—7028060-70
400 ci 350 hp manual—7029263
400 ci 350 hp automatic—7029268
400 ci 366/370 hp manual—7029273
400 ci 366/370 hp automatic—7029270

Distributors
400 ci 265 hp—1111940
400 ci 350/366 hp manual—1111952

400 ci 350/366 hp automatic—1111946
400 ci 370 hp—1111941

Cylinder Head Casting Numbers
400 ci 265 hp—45
400 ci 350 hp manual—48
400 ci 350 hp automatic—16
400 ci 350 hp automatic w/AC—62
400 ci 366 hp manual—48
400 ci 366 hp automatic—16
400 ci 370 hp—722

Option Order Codes and Retail Prices

2 dr hardtop	$3,156.00
2 dr convertible	3,382.00
321 Basic Group	89.53
322 Protection Group	
Hardtop	55.22
Convertible	34.16
324 Decor Group	25.28
331 Mirror Group	
Hardtop	13.69
Convertible	9.48
332 The Judge	337.02
346 400 ci 2 bbl engine w/351	NC
347 400 ci 4 bbl engine	
Ram Air	342.29
Ram Air IV	558.20
348 400 ci 4 bbl Ram Air III engine	168.51
351 Turbo Hydra-matic automatic transmission	227.04
354 4 speed manual transmission	184.80
358 Close ratio 4 speed manual transmission	184.80
359 Turbo Hydra-matic	205.92
361 Safe-T-Track differential	
Exc HD	42.13
HD	63.19
362 Axle, special order	2.11
382 AM push-button radio w/manual antenna	61.09

Option Order Codes and Retail Prices

383 AM push-button radio w/electric antenna	90.58
384 AM/FM push-button radio w/manual antenna	133.76
385 AM/FM push-button radio w/electric antenna	163.25
388 AM/FM stereo radio w/manual antenna	239.08
389 AM/FM stereo radio w/electric antenna	270.68
391 Sepera-Phonic rear speaker	15.80
392 Verbra-Phonic rear speaker	52.66
394 Stereo tape player	133.76
402 Spare wheel tire cover	5.27
404 Rear window defogger	22.12
412 Door edge guards	6.24
414 Concealed headlamps	52.66
421 RH visor vanity mirror	2.11
422 RH & LH visor vanity mirrors	4.21
424 Remote control LH outside mirror	10.53
431 Custom front & rear seatbelts	12.64
432 Rear shoulder belts, exc convertible	26.33
441 Cruise Control	57.93
444 Rally gauge cluster & tachometer (NA w/442 or 484)	84.26
451 Deluxe wheel discs	21.06
452 Wire wheel discs	73.72
454 Rally II wheels	84.26
458 Custom wheel discs	41.07
462 Custom Sports steering wheel	34.76
471 Hood-mounted tachometer	63.19
472 Console	55.82
W/manual transmission	52.66
W/automatic transmission	71.62
474 Electric clock	18.96
482 Tailpipe extensions	21.06
484 Rally gauge cluster & clock	50.55
492 Remote control deck lid	14.74
494 Rally stripes	13.69
501 Power steering	100.05
502 Power brakes	42.13
504 Tilt steering wheel	45.29
511 Front power disc brakes	64.26

514 HD 7 blade fan & clutch	5.27
521 Disc brakes, front only	63.19
524 Gearshift knob, w/floor shift	4.21
531 Soft Ray glass, all windows	36.86
532 Soft Ray glass, windshield only	25.28
551 Power windows	100.05
564 Power seat, LH bucket only	69.51
571 RH & LH head restraints	16.85
578 RH reclining bucket seat w/RH & LH head restraints	84.26
582 Custom AC	375.99
588 Vent, Power Flo	42.13
621 Ride & Handling Package	4.21
622 HD springs & shocks	4.21
631 1 pair front floor mats	4.94
632 1 pair rear floor mats	4.94
634 Superlift rear shocks	42.13
651 Cornering lamps	33.70
652 Luggage lamp	3.16
671 Underhood lamp	4.21
672 Ignition switch lamp	2.11
692 HD 55 amp alternator	15.80
701 Radiator, HD	14.74
702 Space Saver spare tire	15.80
731 Dual-stage air cleaner	9.48
732 Rear compartment throw mat	8.43
754 Front shoulder belts	
W/431	26.33
Wo/431	23.17
Cordova (vinyl) top	100.05
Solid paint, special color	83.20
Two-tone paint, standard colors	40.02
Two-tone paint, special colors	123.22

Exterior Colors*

Starlight Black	A	10†
Expresso Brown	B	61‡
Cameo White	C	50†
Warwick Blue	D	53†
Liberty Blue	E	51†
Windward Blue	F	87
Antique Gold	G	65‡
Limelight Green	H	59§

Exterior Colors

Castillian Bronze	J	89
Crystal Turquoise	K	55†
Claret Red	L	86
Midnight Green	M	57§
Burgundy	N	67‡
Palladium Silver	P	69†
Verdoro Green	Q	73§
Matador Red	R	52‡
Champagne	S	63‡
Carousel Red	T	72†
Nocturne Blue	V	88
Goldenrod Yellow	W	76
Mayfair Maize	Y	40§

*Two-digit code indicates underhood data plate code
†Standard The Judge side stripe colors: yellow-red-blue
‡Standard The Judge side stripe colors: white-red-black
§Standard The Judge side stripe colors: white-yellow-green

Interior Colors

	Buckets	Bench
Blue	250	—
Gold	252	—
Red	254	—
Green	256	—
Parchment	257	267
Black	258	268

Convertible Top Colors

Ivory-White	1
Black	2
Dark Blue	3
Dark Green	9

Vinyl Top Colors

Black	2
Dark Blue	3
Parchment	5
Dark Fawn	8
Dark Green	9

Facts

Following the two-year restyle cycle, 1969 was an off year and the GTO got only minor changes to distinguish it from the 1968 models. As before, it was available as a hardtop coupe and a convertible.

In terms of styling, the GTO retained the Endura front bumper and no optional chrome bumper was available. The Pontiac triangular crest, previously on the top of the nose section of the Endura bumper, was deleted. The grille pattern was changed to an egg-crate design, and the GTO grille emblem was now located on the lower half of the grille. The hidden headlamp option continued to be available. Although the same hood was used, Ram Air decals were placed on the scoops of Ram Air III equipped cars and Ram Air IV decals were placed on the scoops of cars with the Ram Air IV engine option. The front fender GTO 6.5 liter crest was replaced by a GTO emblem that used white letters mounted on a bar that matched the car's exterior color. On the rear quarter panels, the familiar GTO emblem was replaced by a side marker light that matched the shape of the GTO 6.5 Liter emblem. The red lens had white GTO lettering on it.

The GTOs did not have side vent windows, in what seems to have been an industrywide departure from side vent windows. The optional Power-Flow ventilation system was used to circulate interior air.

A GTO emblem was used on the rear deck lid. The GTOs used a different rear bumper that sloped inward at the bottom. A dark gray mat was used in the trunk, and the Space Saver spare tire was once again optional.

The interior followed the same basic pattern as that of the 1968 GTO, but the instrument panel was covered with more padding and the door panels and seats got a different design. Two Rally clusters were once again available: RPO 484 got the Rally clock in the right-hand pod and RPO 444 came with an 8000 rpm tachometer. The standard GTO Deluxe steering wheel had three spokes, which were color keyed to the interior. The optional Custom Sports steering wheel also had three spokes, but only the center hub was color keyed to the interior. The tilt wheel was optional.

The Hurst shifter was once again used with all manual transmissions, but the Hurst Dual Gate shifter for the automatic was replaced with the Rally Sport shifter if the console was ordered. An optional walnut shift knob was available with the Hurst shifters and used a Pontiac emblem rather than a shift pattern.

Early production GTOs got a GTO nameplate above the heater controls if they did not have air conditioning; if they did have air conditioning, the emblem was displaced by a vent. The GTO crest was used on each door panel.

The standard GTO engine was the 400 ci V-8 rated at 350 hp, with the 400 ci 265 hp V-8 a no-cost option. The 265 hp V-8 came only with the Turbo Hydramatic automatic. The Ram Air III engine, rated at 366 hp, was optional. It used D-port cylinder heads and free-flowing exhaust manifolds, and it included a factory Ram Air system. Unlike the hood scoops on previous systems, the ones on this system were functional and the driver was able to open and close them by means of a dash-mounted cable. The Ram Air IV engine, rated at 370 hp, was the ultimate performance engine. It used different heads with round exhaust ports, 1.65:1 rockers, a more radical camshaft and an aluminum intake manifold with a Rochester Quadrajet carburetor and was available only with 3.9:1 or 4.33:1 axle gears. Air conditioning was not available with the Ram Air IV.

All GTOs came with 14 × 6 in. wheels and a hubcap. Optional were the Deluxe wheel covers, which had six cooling slots; the Custom wheel covers, which had five spokes; and the wire wheel covers, which were unchanged. The Rally I styled steel wheels were no longer available; only the Rally II wheels were offered, and these were unchanged. Standard tires were G78 × 14 Red Lines, with whitewalls a no-cost option. Optional tires included fiberglass-belted G78 × 14s and fiberglass-belted G70 × 14s in either Red Lines or whitewalls.

The Judge option was new for 1969, and it was available on both body styles. Visually, it came with a large 60 in. rear deck wing, Rally II wheels without the trim ring, G70×14 belted blackwall tires, special tri-colored side stripes, The Judge decals on the front of the front fenders and on the top right side of the rear wing, and a blacked-out front grille. In the interior, a The Judge decal was placed on the glovebox door; however, the first 2,000 or so Judges do not have this decal. Judges also came with a Hurst T shifter handle. Most Judges were painted Carousel Red.

Carousel Red Judges initially came with red-blue-yellow side stripes, but in late February 1969, the stripe colors were changed to black-red-yellow. Other exterior colors came with different stripe color combinations, but it was possible for the customer to order a Judge with side stripes of his or her own choosing.

All Judges came with heavier trunk lid torque rods to compensate for the heavier deck lid.

The standard Judge engine was the Ram Air III, with the Ram Air IV optional. All regular GTO options were available.

1969 Pontiac GTO The Judge.

Chapter 7

1970 GTO

Production

By Body Style
2 dr hardtop	32,731
2 dr convertible	3,621
2 dr The Judge hardtop	3,635
2 dr The Judge convertible	162
Total	40,149

By Engine and Transmission

Hardtop & The Judge Hardtop
400 ci 350 hp manual	9,348
400 ci 350 hp automatic	18,148
400 ci 366 hp manual	3,054
400 ci 366 hp automatic	1,302
400 ci 370 hp manual	627
400 ci 370 hp automatic	140
455 ci 360 hp manual	1,761
455 ci 360 hp automatic	1,986
Total	36,366

Convertible & The Judge Convertible
400 ci 350 hp manual	887
400 ci 350 hp automatic	2,173
400 ci 366 hp manual	174
400 ci 366 hp automatic	114
400 ci 370 hp manual	24
400 ci 370 hp automatic	13
455 ci 360 hp manual	158
455 ci 360 hp automatic	241
Total	3,784

The Judge
400 ci 366 hp manual	2,380
400 ci 366 hp automatic	1,003
400 ci 370 hp manual	325

400 ci 370 hp automatic 72
455 ci 360 hp automatic hardtop 14
455 ci 360 hp automatic convertible 3
Total 3,797

Serial Numbers

Description
242370P100001
24237—Model number (24237–2 dr hardtop, 24267–2 dr convertible)
0—Last digit of model year (0–1970)
P—Assembly plant (A–Atlanta, B–Baltimore, G–Framingham, P–Pontiac, R–Arlington, Z–Freemont)
100001—Consecutive sequence number

Location
On plate attached to driver's side of dash, visible through the windshield.

Engine Block Codes
WT—400 ci V-8 4 bbl 350 hp manual
YS—400 ci V-8 4 bbl 350 hp automatic
WS—400 ci V-8 4 bbl 366 hp manual Ram Air III
YZ—400 ci V-8 4 bbl 366 hp automatic Ram Air III
WW—400 ci V-8 4 bbl 370 hp manual Ram Air IV
XP—400 ci V-8 4 bbl 370 hp automatic Ram Air IV
WA—455 ci V-8 4 bbl 360 hp manual
YC—455 ci V-8 4 bbl 360 hp automatic

Carburetors
400 ci 350 hp manual—7040263; 7040563 w/EES
400 ci 350 hp automatic—7040264; 7040564 w/EES
400 ci 366 hp manual—7040273; 7040573 w/EES
400 ci 366 hp automatic—7040270; 7040570 w/EES
400 ci 370 hp manual—7040273; 7040573 w/EES
400 ci 370 hp automatic—7040270; 7040570 w/EES
455 ci 360 hp manual—7040267; 7040567 w/EES
455 ci 360 hp automatic—7020268; 7040568 w/EES

Distributors
400 ci 350 hp manual—1111176
400 ci 350 hp automatic—1111148
400 ci 366 hp manual—1112010
400 ci 366 hp automatic—1112009
400 ci 370 hp—1112011
455 ci 360 hp manual—1112012
455 ci 360 hp automatic—1111105

Cylinder Head Casting Numbers
400 ci 350 hp manual—12
400 ci 350 hp automatic—13
400 ci 366 hp—12
400 ci 370 hp—614
455 ci 360 hp—64

Option Order Codes and Retail Prices
2 dr hardtop	$3,267.00
2 dr convertible	3,492.00
Hardtop	
Convertible	
324 Decor group	36.86
Hardtop	
332 The Judge	337.02
344 455 ci 4 bbl engine	57.93
347 400 ci 4 bbl Ram Air IV engine	558.20
348 400 ci 4 bbl Ram Air III engine	168.51
351 Turbo Hydra-matic automatic transmission	227.04
354 4 speed manual transmission	184.80
358 Close ratio 4 speed manual transmission	184.80
361 Safe-T-Track differential	42.13
392 Verbra-Phonic rear speaker	
394 Stereo tape player	
401 AM push-button radio	
W/manual antenna	61.09
W/electric antenna	
402 AM/FM push-button radio w/manual antenna	133.76
404 AM/FM stereo radio w/manual antenna	239.08
411 Speaker, rear	15.80

Code	Description	Price
412	Door edge guards	
421	RH visor vanity mirror	2.00
422	RH & LH visor vanity mirror	
424	Mirror, remote	10.53
432	Rear shoulder belts, exc convertible	
434	Mirrors, outside body color	26.33
441	Mirror, visor vanity	2.11
444	Remote control LH outside mirror	10.53
451	Seat belts, Custom	12.64
462	Custom Sports steering wheel	34.76
471	Deluxe wheel discs	21.06
472	Custom wheel discs	41.07
473	Wire wheel discs	73.72
474	Rally II wheels	84.26
481	Cruise control	57.93
484	Gauges, tachometer, clock	84.26
488	Rally Gauge Cluster	50.55
491	Hood-mounted tachometer	63.19
492	Electric clock	18.96
494	Console	71.62
501	Power steering	105.32
502	Power disc brakes, front	64.26
504	Tilt steering wheel	46.00
514	HD 7 blade fan & clutch	
531	Soft Ray glass, all windows	36.86
532	Soft Ray glass, windshield only	26.33
534	Electric rear window defroster	63.19
541	Rear window defogger	26.33
543	Rear window defroster	53.00
551	Power windows	105.32
552	Power door locks	45.29
554	Remote trunk lid release	14.74
564	Power seat, LH bucket only	73.00
582	Custom AC	375.99
601	Air inlet hood	84.26
611	Driver-controlled Tiger button exhaust	63.00
614	Knob, gear shift	5.27
621	Ride & Handling Package	4.21
631	1 pair front floor mats	6.85
632	1 pair rear floor mats	6.85
652	Luggage lamp	3.16
671	Underhood lamp	4.21

Option Order Codes and Retail Prices

672 Headlight delay	13.00
692 HD 55 amp alternator Cordova (vinyl) top	100.05
694 Moldings, wheel opening	15.80

Exterior Colors*

Starlight Black	A	19†
Palamino Copper	B	63‡
Polar White	C	10†
Bermuda Blue	D	25§
Atoll Blue	E	28§
Lucerne Blue	F	26
Baja Gold	G	55‡
Palisade Green	H	45‡
Castillian Bronze	J	67
Mint Turquoise	K	34§
Keylime Green	L	43
Pepper Green	M	48‡
Burgundy	N	78‖
Palladium Silver	P	14†
Verdoro Green	Q	47‡
Cardinal Red	R	75‖
Coronado Gold	S	53
Orbit Orange	T	60§
Carousel Red	V	65
Goldenrod Yellow	W	51
Sierra Yellow	Y	50‡
Granada Gold	Z	58‡

*Two-digit code indicates underhood data plate code
†Standard The Judge side stripe colors: yellow-blue-red
‡Standard The Judge side stripe colors: green-yellow-white
§Standard The Judge side stripe colors: blue-orange-pink
‖Standard The Judge side stripe colors: yellow-black-red

Interior Colors

	Buckets	Bench
Blue	250	—

Gold	252	—
Brown	253*	—
Red	254	—
Green	256	—
Saddle	255	—
Sandalwood	257	267
Black	258	268

*Not available with convertible

Convertible Top Colors
White	1
Black	2
Sandalwood	5
Dark Gold	7

Vinyl Top Colors
White	1
Black	2
Sandalwood	5
Dark Gold	7
Dark Green	9

Facts

The 1970 was facelifted—although the split grille theme continued, along with the Endura bumper. The GTO looked a lot like the Firebird except that it had four headlights. On the sides, creases above each wheelwell emphasized the muscular look. The rear bumper wrapped around the rear fenders, with the taillight lenses following the bumper's contour and thereby serving as the side marker lamp, as well.

A GTO emblem was used on the grille, along with red, black or white (depending on body color) decals on the rear deck lid and on each front fender behind the wheelwell. A small 455 cid decal was placed beneath the GTO fender decal, if the car was so equipped.

The trunk was finished in a black, gray and aqua multifleck paint and came with a dark gray mat. If the optional Space Saver spare tire was ordered, it was relocated to the left side of the trunk and was mounted vertically.

The interior, except for different trim and appointments, was a carry-over, although a 140 mph speedometer was used for the first time. The top two thirds of the instrument panel was finished with simulated wood appliqué, and the bottom one third was covered with an engine-turned aluminum appliqué. A bench seat was still on the option list.

The Deluxe steering wheel was standard equipment on the GTO. It was unchanged from 1969. Also unchanged was the Custom Sports steering wheel. A third wheel was added, the Formula wheel, which was first seen on the 1969 Firebird Trans Am. The three-spoke wheel was covered with urethane foam. The Formula was available only in black and came with a black steering column. Power steering was mandatory with it.

Power door locks were optional.

The standard antenna on the GTO was imbedded in the windshield. A power antenna was optional and mounted on the right rear quarter.

The engine line-up changed. The 400 ci two-barrel V-8 was dropped and a 455 ci four-barrel V-8 was added to the option list. The 455 was rated at 360 hp with 500 lb-ft of torque. The standard engine was the 400 ci V-8 rated at 350 hp, with the Ram Air III and IV optional. Compression ratios were reduced slightly on all engines—to 10.25:1 on the standard 400 and 10.5:1 on the Ram Air III and IV. Both Ram Air engines came with functional hood scoops.

Transmission availability was unchanged.

All engines came with chrome valve covers, a chrome oil filler cap and a chrome air cleaner lid, except for the Ram Air III and IV engines, which used a black air cleaner lid. The standard 350 hp engine came with a dual-snorkel air cleaner assembly.

The suspension was outfitted with a larger front anti-sway bar measuring $1^{1}/_{8}$ in., and for the first time, a rear bar measuring $^{7}/_{8}$ in. was used. (Other GM intermediates had used a rear bar since 1965.) The steering system was also changed. If power steering was ordered, a variable-ratio system was used with a ratio of 16:1 to 12.4:1.

Because of the 455's increased torque output, GTOs so equipped came with a stronger twelve-bolt rear axle. All others got the ten-bolt rear.

An interesting option was the vacuum-operated exhaust system. When a dash-mounted knob was pulled, engine vacuum would open baffles in the mufflers, thereby relieving back pressure and so producing more power. All GTOs came with a dual-exhaust system with chrome quad outlets.

Wheel size remained at 14 × 6 in., with hubcaps and wheel cover options being carried over from 1969. The Rally II's center cap was changed midway through the year—it was red with black PMD lettering— and the trim ring was also modified slightly, as it had a rounded edge. Standard tires were belted G70 × 14s.

The Judge continued as an option package. The rear wing was redesigned and a small front "chin" spoiler was added. However, the wing was available as an option on other GTOs, and not all Judges came with the front spoiler. The wing was painted to match the body color, though some came through painted black and some came with a small side stripe. The Judge came with a blacked-out grille, blacked-out hood scoop openings, tricolored side stripes over the side creases, and The Judge decals on the front fenders and on the deck lid, replacing the standard GTO decal. The Judge side stripes were also available on regular GTOs. The Judge decals matched the color scheme of the side stripes. A Judge decal was also placed on the glovebox door.

The Judge came with a Hurst T-handle shifter if ordered with a manual transmission. The standard engine was the Ram Air III, with the Ram Air IV optional. Late in the model year, the 455 ci V-8 could be ordered with The Judge.

The Rally II wheels were standard equipment on The Judge, without the trim ring. Trim rings were made available on 1970 Judges.

1970 Pontiac GTO The Judge.

Chapter 8

1971 GTO

Production

By Body Style
2 dr hardtop	9,497
2 dr The Judge hardtop	357
2 dr convertible	661
2 dr The Judge convertible	17
Total	10,532

By Engine and Transmission

Hardtop & The Judge Hardtop
400 ci 300 hp manual	2,011
400 ci 300 hp automatic	6,421
455 ci 325 hp automatic	534
455 ci 335 hp manual HO	476
455 ci 335 hp automatic HO	412
Total	9,854

Convertible & The Judge Convertible
400 ci 300 hp manual	79
400 ci 300 hp automatic	508
455 ci 325 hp automatic	43
455 ci 335 hp manual HO	21
455 ci 335 hp automatic HO	27
Total	678

Serial Numbers

Description
242371P100001

24237—Model number (24237–2 dr hardtop, 24267–2 dr convertible)

1—Last digit of model year (1–1971)

P—Assembly plant (A–Atlanta, B–Baltimore, G–Framingham, P–Pontiac, R–Arlington, Z–Freemont)

100001—Consecutive sequence number

Location
On plate attached to driver's side of dash, visible through the windshield.

Engine Block Codes
WT—400 ci V-8 4 bbl 300 hp 3 speed manual
WK—400 ci V-8 4 bbl 300 hp 4 speed manual
YS—400 ci V-8 4 bbl 300 hp automatic
YC—455 ci V-8 4 bbl 325 hp automatic
WL—455 ci V-8 4 bbl 335 hp 3 speed manual HO
WC—455 ci V-8 4 bbl 335 hp 4 speed manual HO
YE—455 ci V-8 4 bbl 335 hp automatic HO

Carburetors
400 ci 300 hp manual—7041263
400 ci 300 hp automatic—7041264
455 ci 325 hp automatic—7041262
455 ci 335 hp manual—7041267; 7041273 (Ram Air)
455 ci 335 hp automatic—7041268; 7041270 (Ram Air)

Distributors
400 ci 300 hp—1112070
455 ci 325 hp—1112072
455 ci 335 hp—1112073 (HO)

Cylinder Head Casting Numbers
400 ci 300 hp—96
455 ci 325 hp—66
455 ci 335 hp—197

Option Order Codes and Retail Prices

2 dr hardtop	$3,446.00
2 dr convertible	3,676.00
332 The Judge	395.00
455 ci engine	58.00
358 455 ci HO engine	
361 Safe-T-Track differential	45.00
401 AM push-button radio	66.35
403 AM/FM radio	139.00
405 AM/FM stereo radio	239.00
411 Rear radio speaker	18.96

421 Remote trunk lid release	15.00
431 Console	61.00
434 Body color outside mirror	26.33
444 Remote control outside mirror	13.00
461 Custom steering wheel	15.80
462 Custom Sports steering wheel	42.00
464 Formula steering wheel	42.00
473 Wire wheel discs	84.00
474 Rally II wheels	89.00
478 Honeycomb wheels	63.00
491 Molding, wheel opening	15.80
494 Moldings, body side	31.60
501 Power steering	115.85
502 Power front disc brakes	70.00
504 Tilt steering wheel	46.00
511 Power brakes	48.00
521 Floor mats, front	7.37
522 Floor mats, rear	7.37
531 Soft Ray glass, all windows	44.00
532 Soft Ray glass, windshield only	30.54
534 Rear window defroster	64.00
541 Rear window defogger	32.00
551 Power windows	116.00
564 Power bucket seat	79.00
582 AC	403.00
711 Cruise control	64.00
714 Clock w/Rally gauges	51.00
722 Electric clock	18.96
Chrome wheelwell opening moldings	35.75
Triaccent stripes & blacked-out grille	73.85
Rally gauges w/hood-mounted tachometer	118.50
Rear deck airfoil & front air dam	150.00
Lamp Group	37.15
Dual body-colored mirrors w/driver's remote	42.00

Exterior Colors*

Starlight Black	A	19‡
Sandalwood	B	61‖
Cameo White	C	11‡
Adriatic Blue	D	24§

Exterior Colors*

Lucerne Blue	F	26§
Limekist Green	H	42†
Tropical Lime	L	43†
Laurentian Green	M	49†
Nordic Silver	P	13‡
Cardinal Red	R	75‖
Castillian Bronze	S	67§
Canyon Copper	T	62§
Bluestone Grey	V	—
Quezal Gold	Y	53‖
Aztec Gold	Z	59

*Two-digit code indicates underhood data plate code
†Standard The Judge side stripe colors: green-yellow-white
‡Standard The Judge side stripe colors: yellow-blue-red
§Standard The Judge side stripe colors: blue-orange-pink
‖Standard The Judge side stripe colors: yellow-black-red

Interior Colors

	Buckets	**Bench**
Blue	261	—
Ivory	262	—
Saddle	263*	—
Sienna	264	—
Jade	266*	—
Sandalwood	267*	277
Black	269	279

*Not available with convertible

Convertible Top Colors

White	1
Black	2
Sandalwood	5
Green	9

Vinyl Top Colors

White	1
Black	2

Sandalwood 5
Brown 7
Green 9

Facts

Like the 1970 GTO, the 1971 came with a redesigned nose. The grille insert, a crisscross mesh design, extended forward in the grille opening. The hood was new and incorporated two large forward-mounted scoops, which were open on all cars. Ram Air decals were used on the side of each scoop on GTOs so equipped. On the side, twin-bar side markers were used on the front of the front fenders.

Body color side mirrors were new.

In the interior, the door panels were redesigned and a grained panel, instead of simulated wood, was used on the instrument panel. A cassette player was added to the option list. Unchanged were the Rally cluster options and instrument locations on the three-pod panel.

Compression ratios were lowered on all GTO engines and were also rated by the SAE net method as well as the previously used gross method. The standard 400 ci V-8 with an 8.2:1 compression ratio was rated at 300 hp gross and 255 hp net. The three-speed manual transmission was standard, with a four-speed optional. Two other engines were optional, both displacing 455 ci. The first 455 was rated at 325 hp gross and 260 hp net, and it could be had only with the Turbo Hydra-matic automatic. The top engine option was the 455 HO rated at 335 hp gross and 310 hp net, which came with a three-speed manual transmission. The cylinder heads featured round exhaust ports, and the intake manifold was cast in aluminum. The cars carried 455 HO decals beneath the GTO decals on the front fenders. Also, a large 455 HO decal was used on the air cleaner lid. All engines came with valve covers painted to match the engine color, and the air cleaner was painted black.

In addition to the Rally II wheels, an aluminum Honeycomb wheel was available. It measured 14 × 7 in. The Honeycomb wheel was also available in a 15 × 7 in. version, as was the steel Rally II wheel.

This was the last year for The Judge. Owing to poor sales, the option was deleted midway through the model year. It was relatively unchanged and included the 455 HO engine, a functional Ram Air system, the rear deck wing, The Judge stripes and decals, a three-speed manual transmission with a Hurst T-shifter and a blacked-out grille. The rear wing was painted to match the body color, but it could be had in black on white cars. A rare Judge option, RPO 604, had The Judge painted in Cameo White with a flat-black rear wing and black side stripes. Fifteen Judges were so equipped. As all 1971 Judges got the 455 HO engine, a 455 HO decal was used on each side of the rear wing. In the interior, an emblem replaced the Judge decal on the glovebox door.

1971 Pontiac GTO convertible. Pontiac Division

Chapter 9

1972 GTO

Production

By Body Style
2 dr coupe	134
2 dr hardtop	5,673
2 dr convertible	1
4 dr wagon	3
Total	5,811

By Engine and Transmission

Coupe
400 ci 250 hp manual	59
400 ci 250 hp automatic	60
455 ci 300 hp automatic	5
455 ci 300 hp manual HO	3
455 ci 300 hp automatic HO	7
Total	134

Hardtop
400 ci 250 hp manual	1,519
400 ci 250 hp automatic	3,284
455 ci 300 hp automatic	235
455 ci 300 hp manual HO	310
455 ci 300 hp automatic HO	325
Total	5,673

Convertible
455 ci 300 hp HO	1

Serial Numbers

Description
2D37T2P100001
2—Pontiac
D—LeMans
37—Body style (27–2 dr coupe, 37–2 dr hardtop)
T—Engine code
2—Last digit of model year (2–1972)

Description
P—Assembly plant (A–Atlanta, B–Baltimore, P–Pontiac, G–Framingham, R–Arlington, Z–Freemont)
100001—Consecutive sequence number

Engine Codes
T—400 ci V-8 4 bbl 250 hp
Y—455 ci V-8 4 bbl 300 hp
X—455 ci V-8 4 bbl 300 hp

Engine Block Codes
WS—400 ci V-8 4 bbl 250 hp 3 speed manual
WK—400 ci V-8 4 bbl 250 hp 4 speed manual
YS, YT—400 ci V-8 4 bbl 250 hp automatic
YC—455 ci V-8 4 bbl 250 hp automatic
WM—455 ci V-8 4 bbl 300 hp manual
YB—455 ci V-8 4 bbl 300 hp automatic

Carburetors
400 ci 250 hp manual—7042263
400 ci 250 hp automatic—7042264, 7042272
455 ci 250 hp automatic—7042262, 7042272
455 ci 300 hp manual—7042273
455 ci 300 hp automatic—7042270

Distributors
400 ci—1112121
455 ci—1112145
455 ci HO—1112126

Cylinder Head Casting Numbers
400 ci—7K3
455 ci—7M5
455 ci HO—7F6

Option Order Codes and Retail Prices
2 dr coupe	$3,108.00
2 dr hardtop	3,237.00
34W 455 ci engine	55.00
34X 455 ci HO engine	101.40
35G 4 speed manual transmission	185.00

35L 3 speed automatic transmission	242.88
334 GTO Performance Package	353.88
361 Safe-T-Track differential	46.34
401 AM push-button radio	63.00
403 AM/FM radio	139.02
405 AM/FM stereo radio	227.00
411 Rear speaker	14.04
412 8 track stereo tape player	127.00
414 Cassette stereo tape player	127.00
421 Remote deck lid release	14.74
431 Console	61.03
434 Body color outside mirrors	26.33
441 Visor vanity mirror	3.00
444 Remote control LH outside mirror	12.00
473 Wire wheel discs	80.00
474 Rally II wheels	85.00
478 Honeycomb wheels	120.00
491 Wheelwell opening moldings	15.80
494 Bodyside moldings	31.60
501 Power steering	115.15
502 Power front disc brakes	67.00
504 Tilt steering wheel	43.00
511 Power brakes	69.51
522 Rear floor mats	6.32
531 Soft Ray glass	
Windshield only	29.00
All windows	43.18
534 Electric rear window defroster	60.00
541 Rear window defogger	30.00
551 Power windows	110.00
554 Power door locks	44.00
561 4 way power seat	75.00
582 AC	407.59
601 Air inlet hood	57.93
611 Rear deck spoiler	45.00
621 Firm springs & shocks	4.21
634 Unit ignition	77.00
654 Custom trim	168.51
661 Dome lamp	13.00
664 Courtesy lamp	4.00
671 Underhood lamp	4.00
672 Ashtray lamp	3.00

Option Order Codes and Retail Prices

674 Glovebox lamp	3.00
681 Dual horns	4.00
692 HD battery	10.00
714 Rally gauges w/tachometer	80.00
718 Clock w/gauges	50.55
721 Decor Group	35.00
722 Electric clock	18.00
724 Handling Package	186.00
SVT Cordova top	100.05

Exterior Colors*

Cameo White	C	11
Adriatic Blue	D	24
Quezal Gold	E	53
Lucerne Blue	F	26
Brittany Beige	G	50
Shadow Gold	H	55
Brasilia Gold	J	57
Springfield Green	L	43
Wilderness Green	M	48
Revere Silver	N	14
Cardinal Red	R	75
Anaconda Gold	S	63
Monarch Yellow	Y	56
Sundance Orange	Z	65

*Two-digit code indicates underhood data plate code

Interior Colors

	Buckets	Bench	Opt Bench
Blue	—	241	—
Green	274	244	254
Beige	—	245	—
White	272	—	252
Black	276	256	—
Pewter	270	—	—
Saddle	—	—	253

Convertible Top Colors

White	1
Black	2
Beige	6

Vinyl Top Colors
White	1
Black	2
Pewter	4
Beige	6
Tan	7

Facts

The GTO became an option package on the LeMans in 1972. It was available on the two-door hardtop and coupe, and according to the GTO Club of America, one convertible and three GTO wagons were known to exist.

The GTO's front end styling combined elements of the 1970 model with the 1971 model, as the grille was recessed. An egg-crate pattern was used on the grille, and it was painted black. On the left grille opening was the familiar GTO nameplate. The Endura bumper was also available on the LeMans, but it was painted gray and used a Pontiac nameplate. On the side was a functional air extractor vent behind the front wheelwells. The hood was the same one used in 1971.

The rear wing was still on the option list. A ducktail rear deck spoiler was scheduled to be released with the GTO, but only a small number were ever installed.

In addition to the front GTO nameplate, GTO decals were located on the rear fenders and on the right rear of the deck lid. If the GTO was equipped with the 455 HO engine, an additional 455 HO decal was located beneath the larger GTO decal. Decals were available in black or white, depending on body color. Some 455 HO equipped GTOs may not have had the 455 HO decal on the deck lid, and some deck lids also had a Pontiac emblem located in the center.

The GTOs came with side exhaust splitters exiting behind the rear wheelwells. These were mounted horizontally rather than vertically.

The standard interior was the regular LeMans interior, which could be upgraded to the LeMans Sport. Above the glovebox door, the name Pontiac was embossed into the dash pad. The three-pod instrument cluster was surrounded with imitation teakwood. Other changes included the substitution of a 120 mph speedometer for the 140 mph unit used in 1971.

The standard steering wheel was a color-keyed two-spoke unit. Optional was the Formula steering wheel. Most of the Formula wheels used came with a solid Pontiac crest in the center, whereas late 1972 steering wheels had an outlined crest.

Only two engines were available: the 400 ci V-8 rated at 250 hp and the optional 455 ci HO rated at 300 hp. The three-speed manual was standard with the 400, whereas the 455 was available only with the four-speed manual or the three-speed automatic.

Wheels were also a carry-over from 1972. The Honeycomb wheel came with a different center cap that had a red Pontiac crest in the center. Also carry-overs were the brakes and the suspension.

1972 Pontiac GTO.

Chapter 10

1973 GTO

Production

By Body Style
2 dr coupe	494
2 dr sport coupe	4,312
Total	4,806

By Engine and Transmission

Coupe
400 ci 230 hp manual	187
400 ci 230 hp automatic	282
455 ci 250 hp automatic	25
Total	494

Sport Coupe
400 ci 230 hp manual	926
400 ci 230 hp automatic	2,867
455 ci 250 hp automatic	519
Total	4,312

Serial Numbers

Description
2D37Y3B100001
2—Pontiac
D—LeMans body style (D–hardtop, F–sport hardtop)
37—Body style (37–2 dr hardtop)
Y—Engine code
3—Last digit of model year (3–1973)
B—Assembly plant (A–Atlanta, B–Baltimore, G–Framingham, P–Pontiac, R–Arlington, Z–Freemont)
100001—Consecutive sequence number

Location
On plate attached to driver's side of dash, visible through the windshield.

Engine Codes
T—400 ci V-8 4 bbl 230 hp
Y—455 ci V-8 4 bbl 250 hp

Engine Block Codes
WS, WF—400 ci V-8 4 bbl 230 hp 3 speed manual
WP, Y6—400 ci V-8 4 bbl 230 hp 4 speed manual
WS, YG—400 ci V-8 4 bbl 230 hp 4 speed manual
ZS—400 ci V-8 4 bbl 230 hp automatic Calif
YY, XX, X5—400 ci V-8 4 bbl 230 hp automatic
Y3—400 ci V-8 4 bbl 230 hp automatic
YT—400 ci V-8 4 bbl 230 hp automatic high altitude
YK, X7—455 ci V-8 4 bbl 250 hp automatic high altitude
YD, XM—455 ci V-8 4 bbl 250 hp automatic high altitude
YC, XE—455 ci V-8 4 bbl 250 hp automatic
YA, XL—455 ci V-8 4 bbl 250 hp automatic
ZC, ZA—455 ci V-8 4 bbl 250 hp automatic Calif

Carburetors
400 ci manual—7043263
400 ci automatic—7043264, 7043266
400 ci automatic—7043266 (Calif)
400 ci automatic—7043274 (high altitude)
455 ci automatic—7043262
455 ci automatic—7043262 (Calif)
455 ci automatic—7043272 (high altitude)

Distributors
400 ci—1112239
400 ci—1112240 (w/XK engine)
400 ci—1112232 (w/YT engine)
400 ci—1112512 (w/XN, XZ, X5 engines)
400 ci—1112812 unit distributor (w/XX engine)
400 ci—1112233 unit distributor
455 ci—1112220

455 ci—1112507 unit distributor (w/XD, X7 engines)
455 ci—1112203 unit distributor

Cylinder Head Casting Number
400 and 455 ci, all—4X

Option Order Codes and Retail Prices

2 dr LeMans	$3,406.00
2 dr LeMans Sport	3,493.00
35W 455 ci engine	57.00
36E 4 speed manual transmission	190.00
36L 3 speed automatic transmission	236.00
341 GTO Performance Package	368.00
342 Handling Package	4.00
371 Safe-T-Track differential	45.00
411 AM push-button radio	65.00
413 AM/FM radio	135.00
415 AM/FM stereo radio	233.00
419 AM/FM stereo radio & tape player	363.00
421 Rear speaker	18.00
431 Console	59.00
434 Body color outside mirror	26.00
441 Visor vanity mirror	3.00
444 Remote control LH outside mirror	12.00
461 Custom Cushion steering wheel	15.00
462 Sport steering wheel	56.00
472 Finned wheel discs	50.00
474 Rally II wheels	87.00
476 Deluxe wheel discs	26.00
478 Honeycomb wheels	123.00
501 Power steering	113.00
504 Tilt steering wheel	44.00
531 Soft Ray glass, all windows	42.00
532 Soft Ray glass, windshield	30.00
534 Rear window defroster	62.00
551 Power windows	75.00
552 Power door locks w/seatback locks	67.00
554 Power door locks	44.00
561 Power bucket seat	103.00
562 Cruise control	62.00
571 Electric sunroof	325.00

Option Order Codes and Retail Prices

572 Manual sunroof	275.00
582 AC	397.00
711 Electric clock	18.00
712 Clock w/Rally gauges	49.00
714 Rally gauges w/clock & tachometer	100.00
732 Rubber front bumper protector	

Exterior Colors*

Cameo White	C	11
Porcelain Blue	D	24
Admiralty Blue	E	29
Regatta Blue	F	26
Desert Sand	H	56
Golden Olive	J	46
Verdant Green	K	42
Slate Green	L	44
Brewster Green	M	08
Florentine Red	S	74
Ascot Silver	V	64
Valencia Gold	Y	60
Burma Brown	Z	68

*Two-digit code indicates underhood data plate code

Interior Colors

	Hardtop Bench	Hardtop Opt Bench	Coupe Bench	Coupe Buckets
Green	264	—	—	—
Beige	265	—	—	—
Blue	—	271	251	251
White	—	272	252	252
Saddle	—	273	253	253
Black	—	276	256	256
Burgundy	—	—	257	257
Chamois	—	—	—	258

Vinyl Top Colors

White	1
Black	2
Beige	3

Chamois	4
Green	5
Dark Burgundy	6
Blue	7

Facts

GM's A-body intermediates were restyled for 1973, and so too was the LeMans. The GTO, as in 1972, was an option package available on the two-door LeMans and two-door LeMans Sport. The baseline LeMans had a rear quarter window, but the Sport model had vertical louvers covering the window.

GTO identification consisted of a GTO emblem on the blacked-out grille, a GTO decal on the rear deck and GTO 400 or GTO 455 decals (depending on engine) on the front fenders behind the wheelwells. All GTOs came with two NASA-style scoops on the hood, which could be made functional if the Ram Air option was ordered. Fewer than ten GTOs were so ordered. In the interior, a small GTO plate was located on each door panel. Bucket seats were standard, with a bench seat optional.

The GTO came with firmer springs and shock absorbers and a 1 in. rear anti-sway bar. Standard brakes were 9.5 in. drums, with 11 in. power front discs optional. All GTOs came with 15×7 in. steel wheels with a crescent-moon-type hubcap and trim ring and $G60 \times 15$ blackwall tires. The Rally II and Honeycomb wheels were optional, as were two wheel covers: the Custom finned and the Deluxe.

Two engines were available on the GTO: a 400 ci V-8 rated at 230 hp and a 455 ci V-8 rated at 250 hp. A three-speed manual with a Hurst shifter was standard on the 400, with the M20 four-speed and Turbo Hydra-matic optional. The 455 could be had only with the Turbo Hydra-matic automatic. The compression ratio was 8:1 on all engines.

Engines are categorized as early (built before March 15, 1973) and late (built after March 15, 1973).

Early engines were painted a darker blue color and differed in their emission equipment.

All GTO engines came with dual exhausts with chrome extensions.

All other LeMans options were available on the GTO.

1973 Pontiac GTO 400. Pontiac Division

Chapter 11

1974 GTO

Production

By Body Style
2 dr hatchback	1,723
2 dr coupe	5,335
Total	7,058

By Engine and Transmission

Hatchback
350 ci 4 bbl V-8 manual	687
350 ci 4 bbl V-8 automatic	1,036
Total	1,723

Coupe
350 ci 4 bbl V-8 manual	2,487
350 ci 4 bbl V-8 automatic	2,848
Total	5,335

Serial Numbers

Description
2Y17N4W100001
2—Pontiac
Y—Ventura
17—Body style (17–2 dr hatchback, 27–2 dr coupe)
N—Engine code
4—Last digit of model year (4–1974)
W—Assembly plant (W–Willow Run, L–Van Nuys)
100001—Consecutive sequence number

Location
On plate attached to driver's side of dash, visible through the windshield.

Engine Code
N—350 ci V-8 4 bbl 200 hp

Engine Block Codes
WP—350 ci V-8 4 bbl 200 hp manual
YP, YS—350 ci V-8 4 bbl 200 hp automatic
ZP—350 ci V-8 4 bbl 200 hp automatic Calif

Carburetors
350 ci manual—7044269
350 ci automatic—7044268
350 ci automatic—7044568 (Calif)

Distributors
350 ci manual—1112856
350 ci automatic—1112822
350 ci automatic—1112237 (Calif)

Cylinder Head Casting Number
350 ci—46

Option Order Codes and Retail Prices
2 dr coupe	$3,173.00
2 dr hatchback	3,313.00
36E 4 speed manual transmission	207.00
36K 3 speed automatic transmission	206.00
342 GTO Performance Package	
Coupe	452.00
Custom coupe	426.00
Hatchback	440.00
Custom hatchback	414.00
371 Safe-T-Track differential	45.00
411 AM push-button radio	65.00
413 AM/FM radio	135.00
415 AM/FM stereo radio	233.00
421 Rear speaker	15.00
431 Console	58.00
434 Body color outside mirror	26.00
444 Remote LH outside mirror	12.00
461 Custom Cushion steering wheel	15.00
462 Sport steering wheel	41.00
474 Rally II wheels	49.00-87.00

476 Deluxe wheel discs	26.00
501 Power steering	104.00
502 Power front disc brakes	71.00
504 Tilt steering wheel	46.00
541 Rear window defogger	33.00
571 Soft Ray glass, all windows	40.00
572 Soft Ray glass, windshield	31.00
582 AC	396.00
711 Electric clock	16.00
714 Clock w/Rally gauges & tachometer	87.00
734 Rubber bumper protectors	24.00
SVT Cordova top	82.00

Exterior Colors*

Cameo White	C	11
Admiralty Blue	E	29
Regatta Blue	F	26
Carmel Beige	G	59
Denver Gold	H	53
Limefire Green	J	46
Gulfmist Aqua	K	36
Fernmist Green	M	40
Pinemist Green	N	49
Buccaneer Red	R	75
Honduras Maroon	S	74
Sunstorm Yellow	T	51
Ascot Silver	V	64
Fire Coral Bronze	W	66
Colonial Gold	Y	55
Crestwood Brown	Z	59

*Two-digit code indicates underhood data plate code

Interior Colors

Standard Interior
Cloth & Vinyl

Black/Red/White	502
Black/White/Green	523

Vinyl

Saddle/Orange	503
Black/White/Red	555

Vinyl
Green/Yellow	574
Black/White/Red (plaid w/red appointments)	502
Black/White/Red (w/red appointments)	552

Optional Vinyl Buckets
White	262
Saddle	565
Green	567
Red	568
Green (white w/green)	567
Red (white w/red)	568
White (white w/red)	262

Custom Split Bench
Cloth & Vinyl
Black	522
Green	564

Vinyl
White	542
Saddle	563
Green	584
Black (white w/black)	522
Green (white w/green)	584
White (white w/green or red)	544

Custom Optional Vinyl Buckets
White	542
Red	550
Saddle	563
Green	584
Red	550
Green (white w/green)	584
White (white w/green or red)	542

Vinyl Top Colors
White	1
Black	2

Beige	3
Russet	4
Green	5
Burgundy	6
Blue	7
Brown	8
Saddle	9
Taupe	0

Facts

The GTO option was no longer available on the intermediate LeMans in 1974 but was offered on two Ventura models instead: a two-door notchback and a two-door hatchback coupe.

The GTO could be identified by a functional shaker hood scoop, similar to the unit found on the Firebird Trans Am; a GTO decal on the right side of the grille panel; a GTO decal on the rear deck or hatch; and GTO decals on each front fender behind the wheelwells. Decal color varied with body color. In the interior, a GTO emblem was placed on each door panel.

Only one engine was available: a 350 ci V-8 rated at 200 hp mated to a three-speed manual transmission. The compression ratio was 7.6:1. Optional were a four-speed manual and the three-speed automatic. Dual exhausts were standard, using a single transverse muffler. The stock exhaust tailpipes exited behind the rear wheels, with chrome exhaust splitters optional.

Standard on the GTO was a split bench seat, with a variety of bench seats and bucket seats optional.

Standard on the GTO were 14 × 6 in. Rally II wheels without trim rings (they were optional). The E70 × 14 belted blackwalls were standard.

Also standard were drum brakes and 0.812 in. front and 0.562 in. rear anti-sway bars.

All other Ventura options were available in conjunction with the GTO option package.

1974 Pontiac GTO 350. Pontiac Division

Appendix

Transmission Identification

Until 1967, transmissions carried a source serial number, which consisted of a plant prefix letter that stood for the transmission plant, a production date and a shift suffix indicating whether the unit was built during the day or night shift. The following were plant prefix letters:

Prefix	Plant	Transmission Type
A	Cleveland	Manual Powerglide
B	Cleveland	Turbo Hydra-matic
C	Cleveland	Powerglide
CA	Hydra-matic	Turbo Hydra-matic
D	Saginaw	Overdrive
E	McKinnon	Powerglide
H	Muncie	3 speed
K	McKinnon	3 speed
L	GM of Canada	Turbo Hydra-matic
M	Muncie	3 speed & overdrive
N	Muncie	4 speed
O	Saginaw	Overdrive
P	Warner Gear	3 & 4 speeds
P	Muncie	4 speed
R	Muncie	4 speed
R	Saginaw	4 speed
S	Muncie	3 speed
S	Saginaw	3 speed
T	Toledo	Powerglide
X	Cleveland	Turbo Hydra-matic
Y	Toledo	Turbo Hydra-matic

For example, C213N would decode to a Cleveland Powerglide built on February 13 during the night shift.

For 1967, the transmission source serial number was changed to include the model year. Rather than using a number to indicate the month, it used a letter, as follows:

A—January	E—May	P—September
B—February	H—June	R—October
C—March	K—July	S—November
D—April	M—August	T—December

For example, N7B08 would decode to a Muncie four-speed, model year 1967, built on February 8.

Turbo Hydra-matic automatics got an additional number. This number contained the model year, a model identification letter, and the build date code. For example, 67A455 would decode to model year 1967, engine or vehicle model A, build date March 31, 1967. Below this number was the regular source serial number. The build date code for the 1967 model year started with the first day of the calendar year in which the car model was introduced, which in this case was January 1, 1966, and continued through the 1967 calendar year, in this case ending on December 31, 1967.

Effective October 21, 1968, another letter was added to the plant prefix number to facilitate the identification of Muncie transmission ratios. The additional letter codes were as follows:

Muncie Three-speed Manual Transmission

Suffix	1st Gear Ratio
A	3.03:1
B	2.42:1

Muncie Four-speed Manual Transmission

Suffix	1st gear ratio
A	2.52:1 Wide Range
B	2.2:1 Close Range
C	2.2:1 Rock Crusher

In addition, beginning in 1968, the last eight numbers of a car's VIN were stamped on the transmission. Automatic transmissions were stamped on the left side near the top forward portion of the housing. Manual transmissions were stamped on the case.

The location varied in subsequent years, but the VIN was stamped on the transmission.

Cowl Data Plates

An important way to identify a GTO is by the cowl tag. Cowl tags were thin sheet metal tags with stamped numbers and letters that were riveted on the left side of the cowl in the engine compartment.

Several types have been used. The first type, used until 1964, contained four pieces of information: style number (model year, series and type), body number, trim number and paint number. The consecutive sequence number was not included on the plate.

Between 1964 and 1967, a different plate was used that was more informative. Line 1 contained the time built code. This consisted of two numbers followed by a letter. The numbers ranged from 01 to 12 and represented the months of the year. The letter (A, B, C, D or E) represented the week of production (first, second, third, fourth or fifth). For example, 03C would decode to the third week of March.

Line 2 started with the model year, represented by the last two digits of the year. The division series and body series must match the first five digits of the car's VIN. The following three letters represented the assembly plant at which the car was built. The final six numbers were the consecutive unit number, which must match the one on the car's VIN.

Line 3 had the car's trim number and color. It included codes for vinyl top colors and convertible top colors, if so equipped. These codes are included in each chapter.

The plate used from 1968 on was slightly redesigned. The time built code was relocated on the third line and was followed by a modular seat code. The modular seat code was the RPO option code for the type of seat the car was equipped with. For example, A41 would indicate a four-way electric control front seat.

Certification Labels

Beginning with 1969, all Pontiac vehicles included a certification label attached on the inside face of the driver's door. The label stated that the vehicle conformed to all applicable safety standards. In addition, it contained the month and year the vehicle was built, along with the VIN.

Axle Codes

1964

The 1964 GTOs used a color-coding system to identify axle ratio. The color was located on the outer end of each axle shaft, and it may also have been stamped on the right side axle tube next to the carrier. Limited-slip axles had an additional color, green, except 3.9:1 axles, which were green only.

2.56:1	Gray
2.78:1	Red
2.93:1	Orange
3.08:1	Yellow
3.23:1	Brown
3.36:1	White
3.55:1	Blue
3.73:1	Pink
3.9:1	—
4.3:1	Black

1965

In 1965, the color codes were replaced with a two-letter code stamped on the right side axle tube next to the carrier.

1st Letter
W—Standard axle
Y—Limited-slip axle
X—Standard axle w/metallic brake pads
Z—Limited-slip axle w/metallic brake pads

2nd Letter
B—2.56:1
C—2.78:1
D—2.93:1
E—3.08:1
F—3.23:1
G—3.36:1
H—3.55:1

K—3.9:1
L—4.33:1

1966-1974

From 1966 on, a two-letter code was stamped on the left or right side axle tube to indicate axle ratio and type.

Ratio	Std Axle	Limited-Slip Axle
1966-68		
2.56:1	WB	YB
2.78:1	WC	YC
2.93:1	WD	YD
3.08:1	WE	YE
3.23:1	WF	YF
3.36:1	WG	YG
3.55:1	WH	YH
3.9:1	WK	ZK (1966 & 1968) YK (1967)
4.33:1	—	ZL (1966 & 1968) YL (1967)
1969		
2.56:1	WB	XB
2.78:1	WC	XC
2.93:1	WD	XD
3.08:1	WE	XE
3.23:1	WF	XF
3.36:1	WG	XG
3.55:1	WH	XH
3.9:1	WK	XK
4.33:1	—	XM
1970		
2.56:1	WB	XB
2.78:1	WC	XC
2.93:1	WD	XD
3.07:1	WT	XT
3.08:1	WE	XE
3.23:1	WF	XF
3.31:1	WU	XU

3.55:1	WH	XH
3.9:1	WK	XK
4.33:1	—	XM

1971

2.56:1	WB	XB
2.78:1	WC	XC
3.07:1	WT	XT
3.08:1	WE	XE
3.23:1	WF	XF
3.31:1	WU	XU
3.55:1	WH	XH
3.9:1	WK	XK
4.33:1	—	XM

1972

2.56:1	WB	XB
2.78:1	WC	XC
3.07:1	WT	XT
3.08:1	WE	XE
3.23:1	WF	XF
3.31:1	WU	XU
3.55:1	WH	XH

1973

3.08:1	GX	GY

1974

2.73:1	JA	JM
3.08:1	JB	JN

Engine Colors

1964-1965
Robin's-Egg Blue   Ditzler 11561

1966-1970
Silver-Blue Metallic   Ditzler 13255

1971-1974
Enamel Paint
Mixing Color   **Quart Formula**
DMR-441   72
DMR-486   148

DMR-411	182
DMR-490	275
DMR-410	385
DMR-400	605
DMR-495	625
DMR-499	1025

Rally I Wheel Color
Silver Textured   Rinshed Mason E28C009; Ditzler DDL-8568

Rally II Wheel Colors
Silver	Ditzler DDL-8568
Charcoal Grey	Ditzler DDL-32947

Wheel Back Color

Color	Ditzler Mixing Formula
475	6 units
490	106 units
400	272 units
487	442 units
415	447 units
476	457 units
491	477 units
495	497 units
499	1,000 units

Rally I Wheel Part Numbers
1965-68 Rally I wheel	9781264
1965-68 trim ring	9781480
1965 center cap	9781249
1966-68 center cap	9785501

Rally II Wheel Specifications

Part	Part Number	Code Stamp	Wheel Size
1967-68 w/disc brakes	9787279	JA	14×6
1967-68 w/drum brakes	9789329	JC	14×6
1969-71 wheel	485454JW, JT, KT	14×6	14×6

Part	Part Number	Code Stamp	Wheel Size
1972 wheel	485454JW, KR	15×7	15×7
1972 wheel	485455KS	14×7	14×7
1973 wheel	525710JJ	15×7	15×7
1974 wheel	492351–	14×7	14×7
1967-71 trim ring	3923522		14×6
1972 trim ring	9796919		14×6
1973-74 trim ring	488378		14×6
1973 trim ring	488379		14×7
1973 trim ring	488370		15×7
1967 center cap	9787941		
1968-70 center cap	9792996		
1971-72 center cap	480301		
1973-74 center cap	488953		

Honeycomb Wheel Specifications

Part	Part Number	Size
1971-73 wheel	483084	14×7
1971-73 wheel	484425	15×7
1971-73 trim ring	483250	
1971 center cap	9791068	
1972-73 center cap	9795568	

Club

GTO Association of America, Inc.
1634 Briarson Drive
Saginaw, MI 48603

Pontiac Intake Manifolds

1964 4 bbl	9770274
1964 3×2 bbl	9775088
1965 4 bbl	9782895
1965 3×2 bbl	9778815
1966 4 bbl	9782895
1966 3×2 bbl	9782898
1967 2 bbl	9784437
1967 4 bbl	9786285
1968 2 bbl	9794233
1968 4 bbl	9794234
1968 Ram Air	9796614

1969 2 bbl	9794233
1969 4 bbl	9794234
1969 Ram Air IV	9796614
1970 4 bbl	9799068
1970 Ram Air IV	9799085
1971 4 bbl	481733
1971 455 HO	483674
1972 4 bbl	485912
1972 455 HO	485640
1973 4 bbl	496295

GTO Alternators

1964
1100676	std (early)	37 amps
1100683	std	37
1100627	w/A/C	55
1100674	w/trans. ign.	60

1965
1100729	std	37
1100728	w/A/C	55
1100726	w/A/C & PS	55
1100727	w/trans. ign.	42
1100702	w/trans. ign. & A/C	60
1100703	w/trans. ign., A/C & PS	60

1966
1100704	std, w/PS or PS & A/C	37
1100737	heavy-duty wo/PS	55
1100700	A/C, A/C & PS, A/C & PS & AIR	55
1100736	std w/AIR	37
1100699	w/trans. ign., w/trans. ign. & PS	42
1100740	w/trans. ign. & A/C & AIR	60
1100702	w/trans. ign. & A/C w/trans. ign. & A/C & PS w/trans ign. & A/C & PS & AIR	60

1967
1100704	std	37
1100700	w/A/C, w/A/C & PS	55

1100768	std w/AIR	37
1100770	w/A/C & AIR	
	w/A/C & PS & AIR	55

1968
1100704	std	37
1100700	heavy-duty	
	w/A/C	
	w/High Output or w/Ram Air	55

1969
1100704	std	37
1100700	heavy-duty, w/A/C, w/Ram Air	55
1100830	w/Control Spark Ignition	61

1970
1100704	std	37
1100700	heavy-duty, w/A/C	
	w/rear window defogger	55
1100895	w/A/C, w/rear window defogger	61

1971
1100927	std	37
1100928	w/AC, w/rear window defogger	55
1101015	heavy-duty, w/455 ci	
	w/A/C & rear window defogger	80

1972
1100927	std	37
1100928	w/A/C, w/rear window defogger	55
1101015	heavy-duty w/400 ci	
	heavy-duty w/455 ci	
	w/A/C & rear window defogger	80

1973
1100927	std	37
1100928	w/A/C, w/rear window defogger	55
1101015	heavy-duty, w/A/C & rear window defogger	80

1974
1100927 std 37
1100928 w/A/C, w/rear window
defogger 55
1101015 heavy-duty, w/A/C & rear
window defogger 80

Motorbooks International

If you have enjoyed this book, other titles from Motorbooks International include:

GTO 1964-1967, by Paul Zazarine in the Motorbooks International Muscle Car Color History series
Pontiac GTO Restoration Guide 1964-1970, by Paul Zazarine & Chuck M. Roberts
The Big "Little GTO" Book, by Albert Drake
How to Restore Your Muscle Car, by Greg Donahue & Paul Zazarine
Illustrated Pontiac Buyer's Guide, by John Gunnell
American Muscle: Muscle Cars From the Otis Chandler Collection, by Randy Leffingwell
Muscle Car Mania: An Advertising Collection 1964-1974, by Mitch Frumkin

Motorbooks International titles are available
through quality bookstores everywhere.
For a free catalog, write or call
Motorbooks International
P.O. Box 1
Osceola, WI 54020
1-800-826-6600